Managing Offshore Development Projects

An Agile Approach

Upadrista Venkatesh

First Edition

Oshawa, Ontario

Managing Offshore Development Projects
by Upadrista Venkatesh

Acquisitions Editor:	Kevin Aguanno
Typesetting:	Peggy LeTrent
Cover Design:	Troy O'Brien
eBook Conversion:	Agustina Baid

Published by:
Multi-Media Publications Inc.
Box 58043, Rosslyn RPO, Oshawa, Ontario, Canada, L1J 8L6

http://www.mmpubs.com/

Published in Canada.
Printed simultaneously in the U.K. and the U.S.A.

Paperback	1-897326-68-8	9781897326688
Adobe PDF eBook	1-897326-69-6	9781897326695
Mobipocket PRC eBook	1-897326-70-X	9781897326701
Microsoft LIT eBook	1-897326-71-8	9781897326725

Table of Contents

Acknowledgements

I have had the opportunity to interact with hundreds of managers over the last few years of my experience in the IT industry. This book, in a way, is an outcome of those interactions. The successes I had, the problems I faced, the extensive research and experiments I have carried out—all have the same common thread and follow a trend that I put forth in this book. I thank all those people who have worked with me to achieve the common goal of bringing these efforts to success.

I am grateful to all the previous and current employers I have worked with who have given me the rich experience to support this project and for the experiences so important to this project. I am thankful to my sponsoring editor from Multi-Media Publications Inc., Kevin Aguanno, for providing me with important advice on the format of the book and managing the project so smoothly. I also appreciate the efforts of the editing team for giving my flow and expressions a facelift and guiding me through the manuscript development process.

I express my special thanks to Gopal Krishna Behara, K. T. R. B. Sharma, G. S. M. R. Murthy and S. B. Chakradhar for being part of the majority of the case studies in this book. Though I have described them as my employees in this book, in reality these were line supervisors who inspired me during past work experiences. I have used their identities as part of a private joke I share with them alone.

Most importantly, I would like to thank my parents (U. Harigopal and U. Varalakshmi) and my wife Vineela for supporting me in this project, for giving up so many weekends and late nights required to put this together, and for cheering me along through every step of writing this book.

CHAPTER ONE

A Study of the Engagement Model— How It All Started

Today, the global atmosphere favors developed organizations or organizations that have proven, specialized skills when it comes to introducing new services or products within their organization. Such organizations are constantly introducing high technology systems that give them a competitive edge. Due to heavy competition, and with a view to having more market share, these organizations offer services for which they are best suited; in addition, they experiment in new areas to grow the busines into areas with less competition. They keep up-to-date on new technology and processes in order to maintain their lead over their competitors since no company, however friendly or collaborative, likes to lag behind in state-of-the-art technology or processes, especially in critical areas. Processes and technologies, however, do not create a distinction between developed and developing organizations. Processes by their very nature are beneficial to organizations for socio-economic development and the leading organizations try to get the best of these.

However, organizations are comprised of people, and process adherence comes from within rather than from organization enhancement. We have used many processes, but how many of us have succeeded in delivering the expected software at the end of the project? How many engaged in this industry today are successful in their niches with a high client satisfaction record? How many research and experimental findings can you remember that fulfill the following conditions?

1. A model that is completely accepted by the executing teams and that satisfies the developers and the executing teams in terms of the process, delivery model, and customer sense of belongingness

2. Development and testing that are intellectually rigorous and academically acceptable to the customer

3. Predictions that are validated by successful results

4. Predicted results that are the actual results of the project

5. Applications that go to the very heart of the customer

In my own experience, such a combination has occurred in only one in ten projects using the common software development life cycle. There would always be at least one factor from those mentioned above that was not completely satisfied. This is what has led to my fascination with researching a different kind of delivery model, structure, team work culture, and the complementary roles of individuals in the engagements. However, the work is more than fascinating;

it is also extremely timely and results driven. For years, the search for a successful and effective model has centered on a search for the right individual.

Now, during one of my assignments where I was positioned as a Client Engagement Manager for one of my organization's very prestigious customers, my Client Technology Manager came to me and said, "Your software is great but the end users are not delighted." The words were not measurable for me. Before I could consider this objection further, I had to evaluate the nature of the demands and issues from the customer perspective. My initial step towards this was to schedule a formal round table discussion to get all the stakeholders (the end user and technology managers) together to understand the core issues.

One of the major concerns that came out of the interview session was that during the project implementation phase, my organization had not responded positively to the changing requirements. The other side of the coin was that the customer had requested changes when the project was in the final stages of build. My managers had expressed unwillingness to implement those changes as part of the release due to fear of the effect on other application dependencies that might jeopardize the complete release.

This left me contemplating the concept of "customer satisfaction." Regardless of what happens, we have to satisfy the customer in order to compete in the industry, or else our competitors will pick up the work.

Customer satisfaction has been a major challenge across all industries trying to survive and be ahead of others in this current trend of globalization. Not only the IT industry but also almost all the major industries are under high competition at different levels to achieve a larger market share by improving customer satisfaction and strongly focusing on the deciding factors of competitive pricing and quality via the use of technology.

In order to compete in the IT market, major IT companies from the West have been exploring the options of building their software products offshore to cut down on cost. This has made offshore software development a hot and attractive topic these days. Although offshore projects have been quite attractive, many of them have failed to yield the expected benefits mainly due to poor management and processes rather than lack of technical expertise.

Offshore project management is quite different from managing domestic projects due to the many challenges that arise when attempting to go offshore. Some of these are described below.

Trust

Whenever a new project is assigned to a vendor, the trust factor is reduced due to the fear of bad products at the end of the assignment. I believe that trust is usually built from face-to-face communication, but in an offshore environment, most of the interaction happens via email, phone, and online chats. Building trust in this situation takes time.

One of the methods that has worked for me is to have my team present in the client's country for a certain period, mainly for understanding the requirements. Although this is the main goal, I am also of the belief that when people mingle with the natives of that country, they begin to understand the culture, work ethics, trust, etc. In fact, this approach has worked for me with several projects that I have done.

Different Time Zones

During some of my project executions, when I have arrived at the office to start my day, I have encountered the key team players at *my* offshore center packing up to go home due to a hectic night shift to complete a deliverable. Imagine having offshore centers in several countries in different time zones, rather than in one. Next, imagine a situation where your onsite business analyst or the customer is not available to provide clarification to the team so they can proceed ahead to complete a deliverable in the next few hours, as intended.

Requirement Misunderstandings

The majority of developers always tend to be polite and hesitate to be forthright about problems or clarifications. I have come across several situations where the developer does not completely understand the requirements but intends to start working on the code with the partially understood requirements. At the end of the release, the customer highlights the incompleteness of the requirement, which cannot be rolled out until the next release, leading to customer dissatisfaction over the release.

Whenever I ask my developers if they understand the requirement, the answer is often "yes." They will say "yes" to almost anything, just to be polite, no matter how farfetched the issue is. This could be frustrating for a customer, since when the module is delivered, it could be very different from what the customer expected.

Some will also say "yes" when they do not understand even a bit of the requirement. Most of them feel embarrassed to ask questions, thinking the client might feel the question being asked is a silly question.

Documentation

We all traditionally think that the larger the documents we create, the more knowledge they will contain but the fact is, in my experience, that the larger the documentation grows, the more we lose understanding of the highlights.

To achieve our goal, we must:

1. Document what is valuable,

2. Make sure that our efforts are focused on what is valuable, and then

3. Deliver something more valuable than before.

Writing good and crisp requirements is a lot like writing good code. To write requirements well, you must also know the language. Most requirements are written in a natural language (French, English, etc.). Natural languages are very powerful but also very complex; developers not trained in composition sometimes have difficulty communicating complex ideas in writing. We do not have space for a full-blown writing lesson here but some guidelines can help. First, use complete sentences for declarative requirements and second, use simple sentences.

Characteristics of Quality Requirements Specifications

- **Complete:** No requirements or necessary information should be missing.

- **Consistent**: Consistent requirements do not conflict with other software requirements or with higher level (system or business) requirements.

16

- **Modifiable**: You must be able to revise the Sofware Requirements Specifications (SRS) when necessary and maintain a history of changes made to each requirement.

- **Traceable**: You should be able to link each software requirement to its source.

Having said all this, what are the best possible ways to avoid these kinds of issues?

Although the answer to this is not a direct one, what I have found in my research on this topic is that no single process can match these expectations. It is a mix of all the best practices and the learning to be carried forward and grouped as a process that will provide the best kind of model to meet customer expectations and not compromise on the execution team's expectations.

The Need for a Methodology that Works: Introduction to the New Agile Software Development Methodologies

Software development processes and methodologies have existed for a long time and are still significantly evolving to keep pace with:

- The latest trends and advancements in software development languages, tools, and techniques

- More sophisticated software and hardware

- Complicated market requirements

- Higher customer expectations

- The desire to win

In order to face these challenges, software development methodologies have been under constant scrutiny and research. These methodologies are striving to be more effective and efficient for successful delivery of projects with high client satisfaction and better productivity factors within specified and tolerable resource limits.

Most of the current (traditional) software development methodologies being followed have been in existence for many years; however, with recent changes, these models are not proving to be useful in keeping up with the expectations of customers. With the shift towards the traditional offshore model and the requirement for faster development cycles, the traditional methodologies do not seem to deliver results to satisfy the modern-day stakeholders.

From my observations and the current industry trends, the Agile development model has gained popularity, is successful in dealing with current-day requirements challenges, and is an effective software development model.

According to the *Agile Manifesto* from the Agile Alliance:

"We are trying to unleash better ways to develop software with a mature model in place that not only satisfies the customer but also the execution teams. Through this work, we have come to value

- Individual motivation over project demands
- Extensive interactions rather than processes and tools
- Working software over comprehensive documentation
- Customer collaboration rather than contract terms and guidelines

- Accepting a change rather than following a plan

- Customer satisfaction rather than delivery

That is, while there is value in the items on the right, we value the items on the left more."

The principles to be followed for any agile implementation are:

- Give the highest priority to customer satisfaction by providing working software at short and regular intervals.

- Welcome any changes, even late in the development cycle. Agile processes harness change for the customer's competitive advantage.

- Foster customer collaboration with the executing teams to achieve a common project goal.

- Build releases around motivated and skilled individuals.

- Give individuals the support they need and trust to get the work done. Most individuals get into a high motivation mode if trusted by their supervisor.

- Institute face–to–face conversations to the extent possible for more efficient and effective output.

- Use working software as the primary measure of progress.

- Pay continuous attention to technical excellence and good design to enhance agility.

- Realize that the best requirements, architecture, design, development and testing come only from an organized and motivated team.

- Realize that the team may not be completely
productive from day one. Give them guidance
towards the defined goal and motivate individuals
for success. Over time, the team will reflect on
how to become more effective, and then tune and
adjust its behavior accordingly.

In essence, Agile emphasizes people and customer
collaboration over process, tools, plans, and contracts.

Now how does the agile process help us to solve
the problems arising out of what we have categorized
under the trust factors—time zone differences in terms
of communication gaps, requirements mismatch, and
comprehensive documentation? We will now discuss each of
these.

The trust factor plays a major role in any relationship
in a managed services model. It is just as important in an
ongoing relationship as it is in any new engagement we initiate
to have a prospect turn into a client.

However, my discussion here will focus on
relationship building with a new customer, in which trust plays
a very crucial role. We may have demonstrated many of our
capabilities and shared many of our success stories with the
customer during a bid but in the end, *trust* comes from actual
results rather than promises.

Agile differs from a normal development life cycle
in that we can gain the customer's confidence during the
very early stages of the engagement itself by using smaller
releases (spans of 2–3 weeks) during which the customers can
have input about our assignments and work priorities. Our
initial deliveries can demonstrate our capabilities for a longer
and better relationship. After all, "*First impressions are the best
impressions.*" If we are successful during the initial stages of

the engagement, it will hold good for a long time. In addition, the onsite/offshore model plays a very important role in how efficiently customers facing executives provide day-to-day status on the progress of the project.

I personally have success stories in which we have implemented the Agile process for some of my prospective new customers. The first few weeks were a test period for us and my teams worked positively towards understanding and implementing the principles of Agile. This ultimately led to high customer satisfaction from which we were able to expand our footprints to other large assignments. After a year, we raised this account to a multi million–dollar engagement.

How many of us disagree with the fact that, in an engagement, *time zone differences* between the customer and the vendor locations have not caused any hurdles in the project? The answer to this definitely would be very few disagreements. With the onsite/offshore model, we have solved many of the issues arising out of such differences. However, at least a small gap of miscommunication and discomfort always exists. Agile helps solves this issue, though not completely. Agile enforces close collaboration between the customer and the offshore teams on a very frequent basis. This avoids delays in getting the required inputs from the customer, since one factor in the agreement to implement Agile for an engagement is continuous availability of the customer for any needed inputs.

During my experience, I had a case in which customer availability around the clock was an issue due to heavy work pressures on my business users. I solved this issue by deploying a business analyst at the customer location who understood the complete business and could act as an alternate for the business users. He was the point of contact for my executing teams during any periods when the business users were unavailable. However, I do not contend that an alternate will suffice for the complete execution of the project without a business user but can ONLY act as a proxy to the business during emergencies. In the end, "customer collaboration and continuous availability" are core principles of Agile.

Requirement volatility has always been a dominating issue for any customer or vendor. This might be due to many factors such as development teams not completely understanding the requirements, customers not being clear on the requirements during the initial stages of a project, miscommunication of the requirements, etc. To address these issues, Agile recommends splitting the larger requirements into shorter user stories and giving those stories that are clear in nature and needing no further changes priority for development. The concept of shorter release cycles reduces the risk of requirement volatility since the small user stories will have very low chances of misinterpretation and we work on stable requirements, postponing the speculative requirements to the later cycles.

In the majority of my Agile engagements, each iteration (requirements through deployment) spanned two to three weeks' duration. This approach does not leave much room for requirements mismatch with the customer's expectations. In those rare cases where the mismatches occurred, they were solved quickly with the customer as we were working in a collaborative environment.

The Agile principles strictly state: "Working software over comprehensive documentation." In our everyday lives, we try to buy the product that has proven quality, rather than the product marketed with many words. The traditional thinking of long documentation for a better product is overruled in this competing market. The strategy should be to build working software rather than write comprehensive documentation, and Agile solves it all. Document what is valuable. Comprehensive documentation does not ensure project success. In fact, it may increase your chance of failure by wasting resources and introducing complexity.

In one of my experiences, my clients documented user stories using the following format:

As a <<user>>, I would like the system to act << desired format>>

- I will log into the system and it should show me the <<Desired page>>

- I will click on the <<desired link>> and it should direct me to <<Desired page>>

- Etc.

The stories captured in this fashion, supported by execution flows, data flow diagrams, or whichever format is more comfortable to the end user and readable to the developer, are sufficient for estimating and prioritizing and thus, for further planning.

Let me recall earlier days of my experience, where my assignment was with one of the major training providers in the Asian Pacific region. My organization had just entered into a contractual agreement with this customer and I had the prerogative to initiate the engagement. Since we were a very young vendor, we had no established track record but we were a worthy solution provider with a strong work force in place. We had a goal to wind up dominating the market in the training space. Our organization's brand had influenced the customer to some extent, but as we all know, trust does not derive automatically from the credibility of the overall brand. Focusing on strategies to improve the chances of successful project delivery will help the customer gain confidence in the chosen service provider.

Another point to add here is that we had chosen the onsite/offshore model for project execution. The use of this model for project execution, while not a panacea, has been accomplished with varying degrees of success, not only by my organization but by many other corporations in many industries. The promise of the same value proposition—a reduction in resource costs—has encouraged many to pursue the transfer of work offshore. Interestingly, the value proposition of cost reduction is also clear.

Therefore, I started the assignment by positioning my best–of–breed business analyst and technology resources at the customer locations and poised my team to work closely with the customer and to kick-start requirements and technical discussions. The stint for these resources at the customer location was minimal; however, the duration they worked at the customer location was the influencing factor for the customer to understand that the larger set of resources deployed in the project were of similar caliber to those present and capable of delivering the desired project goals.

The other factor that was of utmost importance to me was to highlight the capabilities of the offshore teams. This was targeted by conducting video conferencing calls between the customer and the executing teams at regular intervals where the smartest of my team members discussed the different aspects of the project. While at the same time, I was present at the customer location to demonstrate my team's capabilities.

Synchronous communications methods were used and work was coordinated in overlapping time zones between my offshore location and the customer location to reduce the onsite/offshore model time distance and improve real–time communication. Since we are discussing the geographical separation of the teams, I would like to draw some attention to this idea. That geographical separation alone is not the only issue to be tackled in the onsite/offshore model, however: there are other dimensions to be worked out including:

- Cultural differences that have to be overcome by cross-cultural training. Eventually have team members visit the customer location to understand more about the differences encompassed in social practices, language, and religious beliefs, where possible.

- Administrative differences that have to be overcome, including accounting, legal, and political associations. This will help ensure the effective functioning of relationships in terms of coordination, control, and trust.

- Economical differences that have to be overcome, such as the differences in availability of indigenous resources (human, natural, etc.)

Having perfect co-ordination in the offshore development center in accordance with the client's satisfaction was also one of the major tasks to be achieved.

With all these different factors put into practice, I set out to initiate the 'delivery' in order to gain the utmost confidence of the customer. Thinking back, however, our initial proposal was given using a Waterfall approach and the deadline for delivering the first set of requirements was far away. Therefore, I had initial discussions with my customer and the executing teams to arrive at a practical conclusion of breaking the functionality into smaller pieces and delivering the first piece of functionality for UAT in a shorter time span. This was well received by all and appreciated by the customer since they would be able to see the working system well before the expected dates. Then, after the agreements were modified, the team rolled up their sleeves and started developing the first set of requirements. At the end of the first development iteration, I had been ceaselessly monitoring the progress of the release and my team had done it all; however, a few pieces of the software did not match the customer requirements and expectations. Though these were treated as defects, the turnaround time to fix these issues was minimal, and we were able to run the next set of UAT tests within the span of one week. This time, the customer was delighted to see a bug–free release with zero UAT defects.

Here is the split of the onsite and the offshore responsibilities for the initial set of requirements:

Tasks accomplished by the onsite team:

- Understanding the client's requirements along with those of the offshore teams

- Directly interacting with the client to get a better idea of their needs and changes in them, if any

- Acting as a mediator between the client and the offshore development centers

- Keeping the customer updated on the progress of the project

- Keeping the offshore teams posted on the customer satisfaction level and the expectations

- Planning and designing the initial steps of the project

- Allocating tasks among the available resources in coordination with the offshore project manager

- Testing the outcome of the project in coordination with the client's team

Tasks accomplished at the offshore development center:

- Understanding the client requirements along with those of the onsite teams

- Posting clarifications on the requirements at any point during the project execution

- Continuing the detailed design started by the onsite team

- Deciding on any specific technological requirements for the project

- Developing software

- Testing before handing over to the onsite team

- Providing continuous technical support

Though we had delivered a very small piece of functionality, I got a glimpse of happiness and trust in my customer that eventually turned into a strong bonding and collaborative relationship. This happened, too, in a very short interval after initiating the engagement. The customer appreciated not only the delivery of a bug–free release, but they were overwhelmed and appreciative to see the engagement model of splitting the requirements into a smaller set of releases. This made us quickly build up their confidence level. Not only were they happy to see part of the working system, but they were also able to correct some of the misapprehensions they had about the requirements after having an early glimpse of the system.

Though we initially started to model the engagement under the regular Waterfall model, over the long run, we changed the model to an iterative development model typically known as an Agile process model; however, a lot of homework went into this decision to consider developing this project under Agile. After coming to the concrete conclusion that this engagement was a good fit for Agile, we proposed the model to the customer. We should add to this discussion that when we decide to bid on work, we should bid on processes that we know about and in which we have an interest or at least we should try to acquire the skills upfront before initiating the assignment. Let the team understand this basic principle when having conversations with the customer and you will see a positive difference in building customer confidence.

Finally, what needs to be derived from my experience is that the earlier you gain the confidence of the customer and overcome the barriers, the quicker you will be able to set the engagement on the path to success and grow the business over time.

I have prepared two simple graphs, as shown in Figure 1.1 and Figure 1.2, which demonstrate a comparison between my experiences initiating a new customer relationship following a Waterfall model and another (as described above) which was executed under the Agile (or iterative) development model.

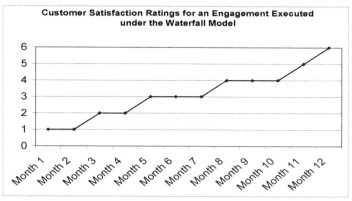

Fig 1.1: Customer satisfaction level during the tenure of the project following a Waterfall model

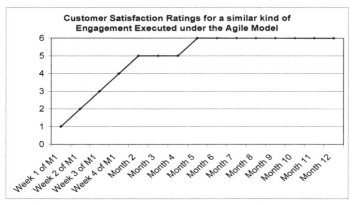

Fig. 1.2: Customer satisfaction level during the tenure of the project following an Agile model.

Both of these engagements were for different customers but the domain and nature of the work, including the size of the assignment, were similar for both of these engagements. The conclusion you can draw from both of these experiences is that with a Waterfall model I was able to gain the customer's confidence after one year working with the customer since the delivery of the product followed the regular sequential delivery model where all the requirements were delivered to the customer for their acceptance testing at the end of the release. Following an iterative approach, I was able to demonstrate to the customer how the system would look in a short duration, which gave the customer more insight into the working model of their requirements using my team's capabilities. This, in turn, helped me to gain the customer's confidence earlier in the game. One point to note is that, in both cases, I was able to gain the confidence of the customer and the only difference was the duration in which this happened. With Agile, we were successful much earlier than in the engagement that used the Waterfall approach.

When I initially started work on the Agile model, I was focusing solely on the principles of effective modeling, but

I quickly learned that the principles cannot be taken as strict guidelines for all engagements. Knowing the concepts behind Agile and applying the knowledge to tailor your processes to the project's needs will lead to a successful implementation. After learning this fact, the majority of my time for any engagement has been spent getting to know the difficult areas in the engagement and applying the relevant Agile principles. This has proven very successful in my career. My implementations on Agile have given me a high client and team satisfaction ratio where my products have been highly accepted and appreciated by the end users without compromising the motivation and appreciation of the executing teams. Though the methodology emphasizes close monitoring and customer availability, the results achieved with this type of model make it time well spent.

CHAPTER TWO

How to Select a Project for Agile Implementation

The development of an approach towards Agile engagements based on the different influencing factors of customer satisfaction and team spirit is a long-running saga. It began for me as a tentative experiment that was narrow in the sense that, as far as was possible, I measured and tightly controlled the inputs and process variables. Since this was my first experiment, I did not want to create havoc with my project and my teams by my experimenting.

Such an ambitious experiment demanded a lot of homework and time. After keen observation on my first successful Agile assignment in the early days, I continued my Agile journey with ongoing experiments and research. My ideas changed and developed over as I tested new Agile program management theories until I gained sufficient understanding into the dynamics of the Agile model to allow successful predictions to be made— predictions that were evaluated against the actual outcomes.

In this chapter, I will be discussing the various factors that can be taken as guidelines to justify considering the Agile methodology for an engagement. Although there are no hard and fast rules to use when considering Agile implementation for a project, the learning and experiences mentioned below will definitely help you to differentiate among the process models to be selected for your project.

Process Evaluator

Over many years, the information technology industry has used many life cycle models that have proven successful. Some of these models are the Waterfall model characterized by the ISO 12207 standard, the Object-Oriented Software Process model and the Rational Unified Process.

Over time, however, these models are proving to be risky in how they deal with customer expectations, indicating that prescriptive processes simply are not fulfilling their promises. Secondly, many developers do not want to adopt prescriptive processes and will find ways to undermine any efforts to adopt them, either consciously or unconsciously. Thirdly, the "big design up front" (BDUF) approaches to software development introduce high levels of risk since they do not readily support change. The reason is that the customers usually are not able to tell you how the entire system should behave up front. They will try, but it is very likely that they will change their minds later in the project. However, in instances in which the customers are very clear from the beginning on what they want to derive from the complete system, then choosing any of the big bang approaches is the right paradigm.

Nevertheless, how many of us agree that the customer provides the complete picture up front? The answer to this might be very few. However, is this the deciding factor for choosing an agile development process, which demands iterations for releases for my project over the big bang approach? This can definitely be one influencing factor in selecting the methodology, but should not be the dominating factor.

Another factor that has influenced me to think of following an Agile methodology in my engagements is that customers almost always have at least a minimal set of requirements mismatch issues at the end of a release. My research had started focusing in this area—into the qualities and attributes of these issues—soon after the basic symptoms were observed and recorded. Since Agile demands the production of working software that has to undergo a pre-UAT cycle by the customer at periodic intervals, this hurdle is overcome. This demand for a better solution to the requirements mismatch problem emerged from the fact that the vendors have not provided the so-called "customer expected software" at the end of the development cycle.

The next factor that influenced me towards Agile was how easily the process can be tailored to a specific project. Many project teams and customers simply adopt the existing process for compliance reasons rather than adapt the process for better results. Agile does not tell us not to follow process; rather, it specifies that we follow processes that will provide the best results for the project and that will be highly appreciated by the customer and the project teams. Process is what guides you out of a storm and the chaos that it produces.

I conceived these three factors as my base criteria when considering whether or not to implement a project in an incremental manner. Having said that, if you select the methodology considering only these factors, your engagement might tend towards failure, as there are bigger challenges that you will face in an Agile development environment. So next I will define the strategies that you have to follow for the smooth and successful execution of any engagement under the Agile process model.

The most important aspect in an Agile–based approach is the availability of the business user to the executing teams. If your business people are too busy with little availability for your engagement, you should not be selecting this model for your project. Agile demands high customer availability. Your business users need to be *part* of the project execution and not just act as the stakeholders for the project.

- Ensure that your business users are always available to provide requirements clarifications. Business analysts can also be effective go-betweens when business users are too busy to provide all the needed availability.

- Ensure that customers elaborate user stories (requirements in Agile terms are defined as "user stories") in a perceptible structure and extend the stories wherever possible.

- Ensure that the customer works to define sets of user stories in order of priority. The analyst teams should have the next set of user story ideas organized before completion of development of the current set.

- Ensure that the customer tests the application after each iteration.

The next principle that helps to determine whether to select an Agile execution model for your project is the availability of your highly-motivated teams as proven resources. Build projects around motivated individuals, then give them the environment and support they need, trusting them to get the job done. Since Agile methodologies usually optimize projects for rapid execution, you should have the smartest and most proven resources within your teams. Nevertheless, I understand that none of us will have all of our best people available for a single engagement, but you need to have at least a high percentage of such people on the team.

Deploy the smartest people you can find. Plan and interview resources on their soft skills, interpersonal skills, and their flexibility in working style without compromising technical or business skills.

Having a contract with your customer is important, but it should not be considered as an alternative to change management. Fixed prices are broken promises. I do not advocate avoiding fixed price engagements for an Agile process model; in fact, I have executed a couple of fixed price engagements based on the Agile approach and they have proven successful. However, for a fixed price engagement, your contract should be defined for each iteration. Select the most critical and well-defined set of functions for the first iteration, and not for the complete engagement. Since the intent of this discussion is not to go into the details of executing Agile on a fixed price model, I will skip discussion about this here.

You can carry out extensive prototype sessions to reduce the risk of change. Again, paper prototyping is a very economical option. If your customer is in agreement with the defined type of contract agreement, this forms a good model for the Agile approach.

Next, if your process demands extensive documentation for project execution, your team will tend to spend more time documenting rather than creating the working software. You should be able to tailor your engagement with as little documentation as is acceptable. If you say that your organization's policies do not permit tailoring the exhaustive tools for process adherence, then an Agile model is not the right fit for your engagement, though most organizations allow for process exceptions with proper management approvals.

Do not mandate the creation of comprehensive documentation during project execution. Reduce the team's time entering huge amounts of data into document templates and reporting tools simply for process adherence.

If you have the courage to overcome all the bare minimum obstacles defined above, then your project qualifies for the Agile process model. However, I will be discussing, over the course of this book, all these principles in detail because the selection of the process model is influenced by the above-mentioned factors.

One of the most misunderstood concepts that I have heard is that success in the first few iterations, first few days or weeks, will determine the success of the engagement; otherwise, the engagement will tend towards failure. This is wrong as there will be initial hurdles that we all have to cross in any engagement. However, what I conceive is that with proper planning and the right expectations set with the customer and the project teams, we can lead the engagement to success from the very first iteration.

Figure 2.1 depicts the typical lifecycle of an Agile project. My terminology here refers to the initial iteration as "Plan" during which we will determine "Initiate," which will be the initial phase of the project. Then, during the project execution, "Cycle 0-A" until "Cycle 0-Z" will be repeated for each iteration. There are some exceptions, based on the project demands, to define whether "Cycle 0-Z" will be repeated for each iteration OR will be executed only once before the closure of the project, which I will be highlighting during the course of this discussion.

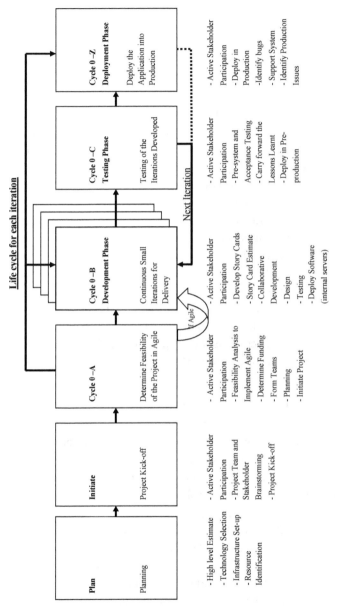

Fig. 2.1: Life cycle of an Agile Project

Planning phase: Though there might be different opinions as to whether the planning phase is part of the project's Software Development Life Cycle (SDLC), how many of us have not revisited the proposal during this time to suggest changes and deviated at least slightly with what has been proposed during the bid? Though this can be part of the bid phase itself, I have separated it for convenience, based on my experience. The first step towards this phase should be the functionality and the right technology identification for the engagement, because technology is the determining factor for the estimate. As an example, an application developed with Microsoft technology will consume less effort than an application developed with Sun Microsystems technology; however, different technical parameters form the basis for arriving at the technology selection for any project. This, however, is outside the scope of this book. The next steps would be to validate the estimates and come up with a high-level plan. The next immediate step should be to identify the right kind of resources for the assignment, and following this would be to get the infrastructure such as the desktops, software installations, network connections, etc. in place

Initiate Phase: This phase will produce a plan and vision for a viable, achievable project that is in line with corporate strategy and overall program needs. Most attention and effort during a project is focused on the end—final delivery. If projects are managed and planned more carefully during the initiation stage, we will have fewer problems in the final stage. To address these aspects, the project kick-off should demonstrate at least the start and the end of the project, the roles and responsibilities of each individual in the engagement, and the project organization structure, including the hierarchy, to show to whom the Project Board reports. Apart from these, you must conduct a feasibility study to investigate the chances of success that you and your project have. This requires a lot of objectivity. It is important to know exactly where you stand and to know your chances of failing the customers' expectations. This phase should also clearly draw a control and reporting regime that is implementable and appropriate to the scale, business risk, and business importance of the project.

Phase "Cycle 0-A": In an engagement based on an Agile methodology, this is the most critical of all the phases. During this phase, you will be determining the feasibility of the project either following an Agile methodology or other life cycle models. There are different factors that need to be considered when selecting a project to be delivered using Agile methods. Some of these are discussed above but we will provide a quick overview again:

- Business user available at all times for collaboration with the teams and performing testing at regular intervals
- Process adherence for the good of the project, rather than for compliance, concentrating more on producing working software
- Skilled and highly-motivated individuals
- Accurate planning for shorter release cycles

These are some of the factors that determine whether your project can effectively use an Agile methodology. Otherwise, you have to select the life cycle model best suited for the assignment. Gaining the commitment of the team members and the stakeholders is only part of the equation. Without the commitment of the customers, functional managers, and other stakeholders and influencers, success will be elusive—decisions will be delayed, clarifications will be postponed, the project will miss deadlines, and politics will undermine the project in subtle ways. Simply stated, all the stakeholders and the project teams, including the technology and business users, should mutually come to an agreement on the acceptance of the Agile terms and conditions.

Phase "Cycle 0-B": Development Phase: Since the project has now gone through the different cycles of reviews and agreements from all the stakeholders of the project, to qualify itself for Agile implementation, the challenging aspect of the project now lies in planning the development phase. The development phase, as I refer to it here, comprises the requirements analysis, design, coding, and unit test until deployment for each function (typically referred to as a *story card* in Agile), and each such cycle will be called an *iteration*. This cycle will be repeated for all story cards until the completion of the final one.

While an iteration might not add enough new functionality to warrant releasing the product, Agile software development methods recommend releasing new software at the end of every iteration. In all Agile cases, it is mandated to release the software at regular intervals of not more than two to three weeks. Regardless, at the end of each iteration, the team should reevaluate project priorities.

Even when used together with an iterative approach, the main focus for success lies in a staged integration. The major risk of big bang integration near the end of development is mitigated with this approach in which integration happens every day, or at least with every iteration. In the traditional Waterfall-oriented model, incremental development has the advantage that you can work in an iterative way only in the background during some activities (coding, integration, testing) and do the up front activities (requirements, analysis and design) up front in a big bang approach. No difficult cultural or process change is needed with your customer. The major disadvantage of this model is that your main approach is to remain inflexible to any requirement changes. With a fully iterative process based on the Agile process model, you have the flexibility to absorb changes in requirements at any point of the life cycle.

Phase "Cycle 0-C": The cycle of testing (typically referred to as user testing–pre-UAT testing) also will follow the iterative approach similar to the development phase, i.e. Cycle 0-B. One most useful fact about the importance of this phase is that rather than trying to track progress by measuring the percentage of completion on intangible elements such as requirements, design, and architecture. Agile tracks progress by fully completed features tested by the users. This avoids the Vaporware Syndrome, where functionality is "mocked up" or features are not developed as per the user-defined guidelines. This avoids the dissatisfaction faced by the customer, when following a Waterfall approach, where the truth—that a lot of additional work is still required to "finish the job"—is not revealed until the end..

As soon as an iteration is ready for user testing, everyone should be on board immediately to perform testing and provide his or her results. Reiterating the point, which I have mentioned earlier—continuous customer availability is a core principle of Agile. Regardless of the method being used, your true goal should be to test and have the system tested by the end users rather than to plan to test, and certainly not to write comprehensive documentation about how you intend to hopefully test at some point. Agilists still do thorough planning, and we still write documents, but our focus is on high-value activities such as actual testing.

Phase "Cycle 0-Z": I hope you have marked "pre-UAT testing" mentioned during Cycle 0-C in your mind. Do you realize why I have emphasized these words? These words have huge relevance in our deployment strategy for the project.

To support the principle I have discussed above in which we had to perform pre-UAT testing on the staging servers, let me elaborate a simple experience of mine wherein we tactfully reduced the cost overruns for one of our major customers.

We were a preferred vendor for a prestigious bank in the United States. We had delivered a couple of projects to this customer and were providing development, support, and maintenance services to these applications. During the course of our relationship with this customer, we initiated Agile for executing the new development projects and were doing well. Apart from my organization, the customer had a huge vendor base, each holding different assignments for different portfolios and business lines. One such business line was staging/production support and maintenance, supported by one of our competitors.

The contract with these support teams states that a cost code is associated with every application deployment to staging or production that the vendor will use to charge the customer for every deployment that the development, application support, or maintenance teams request.

With part of the major agenda of Agile being continuous integration, deployment, and having cost associated with each release for deployment, how do we now overcome this hurdle of avoiding the extra cost on continuous deployment of the working software on the servers? Here is a simple solution that I have applied.

All my iterations were deployed on my integration servers (having made my integration servers replicas of the staging environment). The users would then perform UAT (typically called pre-UAT) after each iteration on these servers.

Finally, after all the iterations were developed and tested on these servers, I moved the complete software onto the staging server and finally to production. Although chances of issues/ bugs evolving within the staging environment persisted, there were very few exceptions. Users performed their acceptance testing on the integration servers, and I refer to this phase as pre-UAT since the customer would accept the final product once they confirmed the software on the staging servers. After the final software was deployed in the customer's staging environment, users would perform a random check of the functionality to confirm the closure of the UAT phase and finally give the go–ahead for the application to be deployed into production.

However, if the customer and organization policies permit the incremental deployment to be carried out on the staging servers, the UAT can be performed in that environment, which will reduce the extra cycle of having two UAT cycles. Even though the final UAT in this approach will involve much less effort from the business users, it is worth avoiding to the extent possible.

You might have noticed in the diagram and during all my discussion that I have mentioned "Active Stakeholder Participation" everywhere as the starting paradigm for each phase. What this implies is that the most significant factor for the Agile development methodology demands continuous stakeholder collaboration so that the right decisions can be made at the right time without any delays.

When the project follows all the principles of Agile, it benefits from a good, healthy energy. Successful managers are able to keep the project in a good spirit. Agile implementation boils down to knowledge, skills, the right processes, and collaboration for getting to the right goal and staying motivated. That's what this book all is about.

Another interesting case study that I will discuss here is an instance where I tried to implement an Agile method in one of my assignments but different influencing factors made me change my mind and not execute this assignment using Agile methods.

It all started in the mid-90s when we were executing a project under the traditional Waterfall model. This was a very privileged customer of mine and we had shared a relationship for the past several years. One day, a customer representative came to me with a proposition for a new project: "The decision has been made—we should be using an Agile development model to kick off my new project. And the command has rung out, 'Make it so!' Now what do you want from me and what will you do to make this happen? How do you take this one requirement of mine and make it happen successfully?"

The idea was well received by me but I had to go back to my desk to think further on the benefits this model would provide to the customer as well as my organization. Therefore, I started my homework thinking on some of the prerequisite factors required to implement the project under Agile. A few of these are listed below:

- **How business-critical is the application?** The answer to this was that we were to develop the application for high-end securities machinery that was highly time critical, and high availability of the application was a mandatory factor. This meant that without having a robust architecture and laid out design up front, the application would not scale to the desired output. In addition, the requirements for the complete application had not yet emerged. By only looking at small pieces (increments) of the software to develop the architecture and design, we might have to do massive rework later in the game. This can be avoided if we analyze or design the whole system correctly up front. This application had reuirements that had not yet been laid out in terms of globalization, performance, etc. A typical example is trying to retrofit "globalization" to an application at the later stages of the project implementation when it was not designed with this requirement in mind. This will lead to lot of rework not only on the application design, but also on the developed code.

- **How critical is thorough documentation of the requirements, architecture, design, and tests?** To develop this application, the project demanded very robust and well-documented requirements, design, and architecture without which the maintenance of the system would be difficult. Ignoring the maintenance of the system, even project execution would be challenged without robust and elaborated documentation, and may lead to failure in the end.

- **Would this *customer* be able to commit the time and effort that the *Agile* methodology demands?** Since the organization was more business centric, all the business users were very busy in their respective areas of work and had very limited time to spend on the project, however important and critical. Agile development wants a customer available throughout the project lifecycle, both to drive the planning process and to assist with requirements refinement issues as they arise. The other factor was that there were too many relationships between the customers and the development teams and it was hard to establish the "customer-always-available" practice with this required interactions.

- **Do my development teams accept the Agile model?** During discussions with my teams, I had determined that the teams did not fully understand the concepts of Agile. One of the reasons was that they were very new to this model and had not seen or heard its benefits. Another was that they were not well trained on Agile. Convincing the teams to adopt an alternative model, however, was definitely one option open to me. I always believe in acceptance rather than enforcement, and acceptance has to come from the heart. Although this was a factor in my judgment, it was not the major influencing factor since having my teams accept the right process in the right spirit had never before been an irresolvable obstacle for me, however complex it might be. I am a great negotiator. Secondly, I do not go to any team making a big deal about trying any new practice; in fact, I try to bring as little additional attention to it to the team as possible. The team will need time to work through even a small change, to measure its benefit, to change its model, and finally to embrace it.

- **Does this engagement really need an Agile and iterative development model in which the customer has to validate the system incrementally?** I started to collect answers to this question by finding out the issues the customers were facing, or had faced in the past when executing similar kinds of projects. I focused on some of the key aspects of whether there were any disputes in the assignments (with the same customer) around requirements mismatch, schedule, or effort overruns. And the answer to this was that we were always positive in all these aspects and had no customer complaints. However, on this project, the demands for continuous availability from the business users were not achievable and other factors also influenced my decision.

At last, I had answers to all the data points listed out which clearly defined the future of this specific engagement. Refer to Figure 2.2 where a simple graph was derived to ascertain the implementation of Agile in this engagement. This graph finally determined that this engagement could not be implemented with Agile principles. A great deal of homework went into this research.

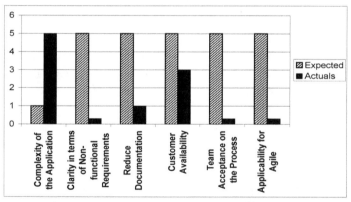

Fig 2.2: Analysis of my engagement dynamics

I finally got back to my customer with all these points for discussion and presented the challenges of executing the project using Agile. I was able to convince them to follow the traditional Waterfall model for this engagement since it was well suited. My counter arguments were well accepted by the customer since I had all the valid data points to support my judgment. I clearly explained the order in which an Agile implementation would be successful and the patterned approach I followed for this engagement.

One change that I have observed in this current market is organizations adopting Agile methods just to add credibility to their brand. Please note that Agile has not always proven successful and determining the right engagement in which to follow Agile is the predominant factor for success. Before we start changing things, let's first step back and be sure of what we want to achieve and understand what it is that we are trying to fix. After all, if there were not a problem with the way we do things today, there would be no reason to change. Therefore, our starting point is the original argument for adopting an Agile method and its applicability; however, many have told me that they are at risk of losing business with a customer because the customer is suddenly concerned that they are not an "Agile" shop.

The other side of this situation is that there are many mature organizations that have not changed their paradigm by branding themselves as "Agile," understanding the fact that not all engagements can follow the Agile methodologies. This is merely because of the nature of the job to be accomplished and they are nonetheless successful and competing well in the market.

Many organizations have also had a little lingo adjustment, naming themselves Agile shops. With some good marketing, they can make short work of differences with their potential competitors. These organizations, however, have not reaped success in the end. As people who call themselves "Agile" create messes and as project disasters become known through the local community, the shine certainly wears off.

With so much hype in the market for Agile, the majority of organizations want to execute their engagements on Agile without understanding the crux of what the methodology really demands from each individual, whether it is the vendor or the customer, and what benefits each can reap from the methodology. One of the things I have observed with many of the individuals around me is the sometimes false perception that we can gain a more quality output by implementing the assignments under the Agile process models. Under many circumstances, I personally have had experiences where Agile methods did not work at all but so many other methodologies did work. And this was the source of my frustration with many individuals who want to gain something without knowing the basis of what needs to be achieved.

For example, should an organization accept Agile as its development model for projects that are highly critical and life-threatening? Not a wise decision. The reason is that the lack of design concepts and documentation for the complete application and the de-emphasis on the architecture are the riskiest parts of these kinds of engagements. Remember, there is no silver bullet. Developing complex and highly-critical applications without knowing the entire system up front will lead the engagement to disaster.

There will be many pitfalls along the way, which cannot be overcome. Use what will work best for your situation. Leverage the skills of the people that will be working on the project. And most of all, do what makes sense. Agile methods can be effective but so can other methods. We must determine why we are embarking on this course, choose the method that will satisfy the need most closely, then map out the path from where we are today to where we need to be. Then we can "make it so."

As more and more software projects adopt Agile methods, there are emerging patterns of success and failure.

CHAPTER THREE

Resourcing

All of us want to be part of a winning team. In a running competition, the gold medalist may have beaten the silver medalist by a mere 0.001 of a second, but we, the great majority of the viewers, treat the gold medalist as superior in the class. Here, the winner's confidence and hard work make the difference. In contrast, a soccer team has a success by the entire team performing with a winning spirit and using a winning strategy. If anyone on the team fails to perform, the entire team loses. Only a motivated and skilled individual can be a winner, displaying the winning spirit within him to others and inspiring the entire team to win.

In today's world of downsizing, doing more with less, and "working smarter, not harder," teamwork is more important than ever. No group of people or individuals working in isolation can do better than several key individuals working together. In fact, one definition of a winning team is one in which "the whole is greater than the sum of the parts." Synergy is essential but elusive. The best team attracts the best people.

In the software industry, success in any engagement is always centered on people and resources. It is not about knowing how to write code but knowing how to write for a better software product. I have personally experienced that it is better to have motivated and skilled people on the teams, communicating and working towards a common goal using little formal process than to have a well-defined process used by unmotivated individuals.

One important object of an Agile project implementation is to gain and sustain commitment to the project's mission. Commitment occurs when the project stakeholders, including the project team, all agree on defined guidelines and the execution model. In addition, buy-in occurs when there is a critical mass of desire to support the execution model, as well as confidence in its ability to succeed. You realize that commitment is present when you have a stable and motivated project team and the support from the stakeholder community at large.

When both these elements are present, the project gains momentum in its journey to successful completion. The project stays in a positive energy despite any initial fallbacks. Keeping the individuals and the project in high spirits is a top priority for a successful Agile engagement. When commitment from individuals vanishes, the project's energy level turns sour, and there are increased chances that the project will turn red. It then runs the risk of not meeting schedule, quality, and often the customer expectations; therefore, to gain and sustain positive energy in an Agile project, identifying the right kind of resources and keeping their momentum high during the tenure of the project execution are primary tasks of an Agile project leader. You want an effective team but also a sustainable one. Otherwise, burnout will result, leading to stress and employees quitting, which in turn lead to failure.

This underlying principle not only holds good for any Agile development model, but holds equal importance for whatever life cycle model you select. The only difference is that, with the Agile model, you will be able to see success or failure at the early stages of project execution, whereas with the Waterfall model, you might only see failure at the end of the project.

Ways to create a winning engagement

Many people think they can sell ice cubes to Eskimos—but truly great resources are rare. To find the crème de la crème, look for resources with these characteristics:

1. Highly motivated

2. Enthusiastic to learn

3. Able to cope with rejection

4. Self–confident

5. Appreciative of a challenge

6. Persistent

7. Competitive

8. Great listening skills

9. Physically and mentally energetic

Hiring the Right Teams

During my early stages in the leadership role, I first realized that being a leader means having to change how you act. Too often, people who are appointed to their first leadership position miss that point and trip up their careers. Being a leader has changed everything for me.

Before entering into a leadership role, my success—my performance, my technical skills, my contributions, and my solutions—centered on the technical area. Once I transformed into a leadership role, success was all about growing my people and making people who work for me smarter, more skilled, and bolder. Nothing that I have done as an individual has mattered as much as how I nurture and support my team and increase their self–confidence. My success as a leader has not resulted from what I do but from the reflected glory of my team.

Fig. 3.1: On the job to victory

So, hire the smartest and highest-motivated leaders for your engagement since Agile demands the smartest and most people–centric leaders to motivate and groom the team for high productivity. Being a leader, you should no longer think of how you stand but think how you can help your teams to do their jobs better. This often requires a mind shift.

Unlike the Newtonian model, where power and success come from position and designation. Leaders in the Agile model will have little direct operative authority over the people. Their power comes from the relationships they establish. In this quantum model, the Agile leader is someone who unleashes power rather than imposes power. As a leader in this style, you can be as good at managing and leading others as you are in managing and directing yourself. If you are constantly overstressed, anxious, lack discipline, or have no clear goals, you are incapable of Agile execution, as Agile enforces very small time boxes for iterations and failures also tend to come in short spans. By your position, you may be able to make others comply, but you will not be motivated or inspired when the chips are down. To be able to manage others effectively, you should be able to manage your emotions effectively and move into the right spirit of the game.

Although the Agile leaders oversee the project's entire energy field, their primary job is to manage the context, that is, facilitate the flow of emotions, thoughts, new ideas, and interaction in a way that will provide the desired outcomes. As Agile leader, you are the project's energy leader.

Now hiring or deploying the workforce is the next challenge for a successful engagement. You should

- Hire the smartest people you can find.

- Plan and interview resources on the soft skills, interpersonal skills and their 'flexibility to work.' Do not compromise on the technical/business skills. Ensure a willingness to work on alternate shifts.

- Assess behavioral skills.

- Assess the right match to your project in terms of technical capabilities.

- Assess the right match to your project in terms of domain capabilities.

- Determine whether people are highly motivated in the organization and can work under pressure and in shifts, if needed. The Agile methodology demands highly motivated individuals who have no constraints on work times, their flexibility to work on shifts, etc.

- Assess his interpersonal skills.

- Assess communication skills since the team needs to interact constantly with the customer during the overall project life cycle.

- Prefer resources who demonstrate a real commitment to continuous learning.

Keep the Team Motivated

Motivating a group of individuals is often more challenging than motivating a single individual. Individuals within teams operate with different goals, values, beliefs, and expectations. Yet the variety of team member personalities can be a positive force if each performer contributes his or her unique capabilities when and where needed. Teamwork potentially allows a number of individuals to achieve more when they collaborate than when they work separately. As a leader, the first principle is that you need to be able to motivate your corporate team. Agile methods demand that the team be highly motivated and focused on work.

Motivation is one of the key psychological aspects needed for a successful journey. In general, at the start of the engagement, all of the teams are highly motivated and have the spirit to win. However, over time, due to the pressures and hurdles in the project, the spirit of the teams slowly decreases. A good leader is the one who keeps up the teams' motivation and spirit under pressure. The first thing a leader needs to understand is that the teams' motivation cannot always be high. The journey is very long and individuals cannot perform to their fullest potential all the time. The leader's responsibility is not to keep motivation high all the time but to understand and implement how to handle the fluctuations in motivation periods in which an "extra boost" in motivation is needed for the benefit of the team performance.

However strong and skilled your teams are, without proper motivation, recognition and their acceptance to do the work from the heart, any engagement can lead to failure if the team's morale is not high.

I remember the old days when I once I visited a hotel along with my brother in New York, where we met an interesting guy named Paul. Paul was solely responsible for the hotel's baking using automated and highly-sophisticated machinery. Paul was quick to point out the system that he was using to monitor the baking processing unit. He not only had the power to control the unit by speeding up or slowing things down whenever required but also had the authority and the power to stop the complete system. I asked him who had designed this system that had given him so much control over the whole process. He told me, and added proudly, "I worked with the design engineers on what I needed to operate this machine and what type of problems I might run into." Paul was a fan of the system as he played a major role in creating the system, though not directly. Still, his ideas and wishes were considered to be of utmost importance. Not only were his hands and mind engaged but also his heart.

In practice, the principle of "ownership of tasks" means to give individuals the freedom to determine how to do their assigned tasks as well as the opportunity to influence how to succeed on the overall engagement. In particular, it means:

- Give individuals the support they need.

- Recognize that people prefer to participate in changing themselves by themselves rather than to be changed by others.

- Give people the voice they always crave so that others can hear them on the issues and topics that are near and dear to them, even if you are the person who will make the final decision.

- Provide guidelines rather than dictating terms as rules.

- Involve people in group meetings.

- Reward the appropriate individual in front of others. This will give others the winning spirit to do more to try and reach the reward category.

This accelerator is based on the fact that people support what they want to create.

Instill the Six Philosophies

1. **Support**: Give the team the environment and support they need, and trust them to get the job done.

2. **Recognition**: Ensure resource willingness to work in alternate shifts. Provide them with the rewards and recognition they deserve for high motivation.

3. **Commitment**: It is essential that you lead the mission and that the entire team follows your example.

4. **Cooperation**: The whole must become greater than the sum of the parts.

5. **Communication:** Provide all necessary information, and let the team members know that it is necessary to share data with each other, fellow colleagues, and the customers.

6. **Contribution**: Participation is not optional in a teamwork situation. You must require it and support it.

Leaders need to ensure that Human Resources creates effective mechanisms such as money, recognition, and training to motivate and retain people. In addition, they must confront strained relationships with unions, individuals who are no longer delivering results, or stars who are becoming problematic.

Some great ideas for team motivation:

• Let your team know that they are wanted and are the core players in the project.

• Greet or acknowledge people each day.

- As people share their ideas, recognize them and use the ideas.

- Constantly strive to avoid work overload. Give your team the right kind of support they need.

- Be responsive—get back to their requests promptly.

- Increase responsibility—everyone needs to grow.

- Give your team credit for a job well done. Do not grab the opportunity for your glory.

- Make sure that the company at large recognizes good work done by your team. Keep your team in the company's eye.

- Look for better careers for your team members. Do not hold them back. Give them training. Encourage them to take responsibility and advance in the company.

- Write letters, notes, and e-mails, or leave voice mail saying, "Great job!"

- Give them public recognition—meetings, certificates or plaques (be specific).

- Make sure top management recognizes them.

Having said all these things, I personally and experientially feel that it is not always practical or feasible to have 100 percent highly motivated and highly skilled resources on the team. In addition, with the upcoming trend towards college hires and organization policies of retaining them in the projects, it is not always possible to have the same level of expertise and energy at each level in a team. So how do we cope with the agility of the project deliveries that demand highly skilled and motivated resources? As I said previously, a leader has to set an example for the teams. He has to motivate and groom the team for high productivity. In my experience, to have a successful Agile implementation, your team needs to be made up of 70 percent *proven*, motivated, and skilled resources, and with the remaining 30 percent you can take a risk. Please note the emphasis on the word "proven." It has huge relevance by itself. These resources should really have proven their skills in their previous experiences. However, never plan for this percentage mix if you know you can get more than 70 percent proven resources on your teams.

A simple definition of a team in terms of the amount of experience and expertise required at each level to achieve common goals is my next illustration. From my experience, I have derived a resource mix at each level that has proven successful for me. Beyond this definition, there are also numerous variations in team composition and structure, but the self-organizing team seems best fitted for exploratory work. In an organized and disciplined team, individuals are assigned to defined and clear roles and responsibilities for managing their own workload, assign shift work among themselves based on need and best fit, and participate in team decision making. Team members have a healthy leeway in how they deliver outcomes, but they are accountable for those results and for working within a defined but flexible framework.

The resource pyramid defined in Figure 3.2 defines the resource mix that you should be adapting for a successful Agile implementation. If you look at the graph carefully, you will notice that I have defined the percentage composition only at the level of roles and have not inferred any distinction in terms of the years of experience for each role. This clearly defines the fact that roles should be determined based on resource skills and the acceptance level of the other team members in the project for an individual. A simple case is one in which a two–year experienced member of my team has taken on

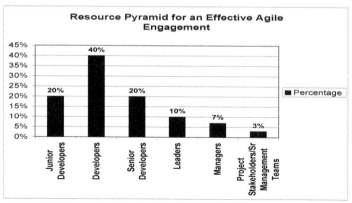

Fig. 3.2: Resource pyramid for an effective agile engagement

the responsibility of a technology leader in one of my Agile engagements and has demonstrated huge success with his module. This model of matching the right set of individuals with the right roles not only provides an opportunity for the skilled resources to be recognized and motivated but also moves the project team towards the right winning spirit.

- **20% Junior Developers**—College graduates who have been trained in the technology and have proven their skills in one form or the other.

- **40% Developers**—These resources should be hands-on resources and should have a grasp of the domain and the technologies.

- **20% Senior Developers**—These resources should be leaders motivating the team and who are providing direction to the junior resources.

- **10% Leaders**—These resources should be those who lead the teams and can make an example of themselves.

- **7% Managers**—These are the people who monitor the engagement and track the status on a periodic basis.

- **3% Senior Management Teams**—These are the top level management who handle escalations.

Our final set of discussions will concentrate on the operation model of the offshore and onsite blend. The offshore model and onsite model are never mutually exclusive and moving a good amount of work to offshore has always been a challenge. Overall, tremendous economies of scale of almost 30 to 50 percent cost savings of an offshore model over a complete onsite model have gained popular recognition. Today, the most appealing and most mature delivery model is onsite–offshore, where the company combines local support with offshore resources.

Successful offshore outsourcing deals combine a mix of confidence, trust, control, openness, and teamwork. The key ingredients of a mature and successful collaboration at work are choosing a compatible, competitive, cooperative environment and blending this with a role–based management team that is motivated to make the arrangement work. What has worked best for me with an Agile–based engagement model is an onsite–offshore ratio of between 30 and 70 percent with no deviation. The most important role in this proposed model is the onsite champion, who acts as the conduit between the offshore team and the customer. The 30 percent or greater blend of onsite teams should include the leaders, coordinators, business analysts, and technology teams, which would cover the complete borders of the project. However, for an ongoing relationship and during the tenure of the engagement, you can facilitate and plan for reductions in the cost of the project by reducing the onsite ratio; however, it should never be a compromise. Agile is demanding and has never succeeded with any compromises.

Another key aspect of adjusting your process to build the appropriate communication channels between the onsite and the offshore teams is to establish the right mode of communication with the right set of individuals assigned to the roles. The D-D model adopts direct communication between people at each level of the project and their counterparts offshore. Onsite analyst teams work directly with the analyst teams offshore, as do the developers, project managers, and program managers. It fosters more of a single-team environment when team members in all locations are able to interact directly with one another. Your onsite partner is an extension of your own team. This model works best but it should never be defined as the protocol. In the end, everyone should be able to reach anyone in case of any needs relevant to that particular individual.

Trust and confidence are achieved with close interaction, regular communication, and face-to-face communication. In addition, you should institute regular teleconferencing with the onsite–offshore teams. Depending on the complexity of the problem and the project, it may need to be daily or every other day. Use e-mail to help clarify topics and get answers to quick questions. The pivotal point in such communication is that it has to be bidirectional with equal vigor and engagement. In addition, travel at all levels by both onsite and offshore teams should be encouraged, so that one understands the criticality of the environment of the other.

It All Worked for Me

The onsite–offshore model for an Agile engagement has worked very well for me and has delivered tremendous business value and customer satisfaction. Furthermore, the team motivation has always been high because everyone is always considered an integral part of the engagement. It has provided me with resource motivation, filled the gaps of cultural differences, provided process improvements, and most importantly, provided cost reductions that have driven major improvements towards budget utilization and financial performance. An increase in the offshore percentage from 45 percent to 70 percent reduces the relative cost of operations by about 30 percent. Short-term outsourcing reaps tactical rewards from cost diminutions obtained through rate arbitrage.

In one of my experiences, where I had implemented the Agile approach with an existing customer of mine, there were huge challenges in terms of team compositions and the resource blends with the existing teams. It is not possible for me to include the full details of the case study herein, but I will provide a summary of how I approached the resourcing issues in the assignment.

This engagement was for a major leasing company in Germany. Each project we had been handling for this customer had a schedule spanning between 8 and 16 months and teams comprised of 50–60 resources at the peak. All projects being executed for this customer were of a similar nature.

One day, the vice president of my customer organization met with his group of program and project managers and me and indicated the organization's desire to use Agile software development for their engagements as "the way forward" in their organization. This was not the beginning of this company's adoption of Agile methodologies, but rather the culmination of more than a year of learning, experimentation, and prototyping that made them want to initiate the engagements under Agile.

Adhering to a big-bang approach to change all the engagements was a risky proposition, and dramatically changing the way a risk-averse, highly-regulated company develops software requires more than just a successful pilot and a couple months of coaching. Therefore, the idea was to pilot a release under Agile.

In addition, based on industrial data from other companies, my customer had done considerable homework to compare the productivity of Agile methods with the conventional approaches we had been following in the previous engagements. We had followed ad-hoc OO software development practices (utilizing use cases, sequence diagrams, UML diagrams, etc.) before adopting Agile techniques the previous year. The customer was very enthusiastic about using the Agile practices in his engagement by switching in lightweight approaches, and I started to understand the proposition.

In this experience, I was able to set all the other parameters for a smooth Agile adoption apart from some of the predominant hurdles I had with the engaged resources. With the current executing model, my teams were able to deliver the right kind of output, as the deliveries spanned over longer durations, and a superior work force was used to balance the weaker workforce. Nevertheless, Agile would have moved the project into an unsuccessful spirit without close supervision and changes to the resourcing model.

Let me summarize the empirical findings from the engagement. I started gathering metrics from my managers for each resource in terms of his or her technical capabilities, number of years of experience, motivation, interpersonal skills, and customer interaction skills and finally derived a consolidated rating for the engagement. I used a rating from 1-4 of which

- 1 stood for not satisfactory,
- 2 stood for satisfactory,
- 3 stood for good and
- 4 stood for very good.

Figure 3.2 demonstrates the results for this engagement.

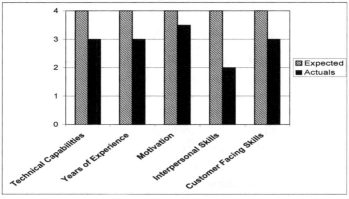

Fig 3.3: Existing resource trend for the engagement

- **Technical capabilities**: On average, the technical capabilities of the resources deployed in the team were not uniform and there was a huge trend towards having team members with superior skills to execute the assignment.

- **Years of Experience**: The resources deployed in the engagement were not uniform and the resource mix with the existing teams (in terms of experience and the roles) was not at a desired level to run an Agile–based engagement.

- **Motivation**: The motivation level on average within the teams was low. This probably was due to some bad leaders. Predominant individuals within the team operated with different goals. I heard that at the start of the engagement each individual was highly motivated and shared a common spirit, but over the long run, the motivation of many individuals went down. This was due to monotonous work, pressure, hurdles in the project, bad leaders to motivate the team, etc. The leaders were not able to align the principles of teamwork and a common goal for all individuals successfully.

- **Interpersonal skills**: The interpersonal skills within the teams were low and a few dominating members of the team were leading the engagement. The main things that affect all personal interactions are speaking, listening, and mood awareness. All these three aspects of the team's personal interactions for effective coordination were lacking.

- **Customer interaction skills**: Dominant members of the team were not exposed to the customer in the ongoing assignment while the leaders of the project collaborated with the customer. This left other members of the team lacking in customer interaction skills or the chance to prove themselves in this area.

The next set of findings revolved around the resource mix for this engagement in which data was collected on the roles of each resource. Figure 3.4 demonstrates the results.

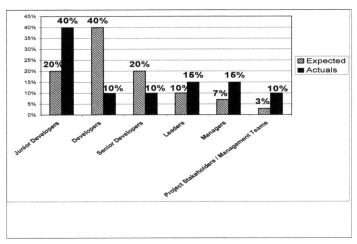

Fig 3.4: Existing resource mix in the engagement

The findings revealed that the leaders were dominant in the mix, with a huge mismatch in terms of the expected number of developers, junior developers, and senior developers to run the assignment. This happened because in an ongoing assignment, any resource requirement was immediately serviced based on the need, and thinking about the bulge ratio was always a low priority. With a traditional life cycle model, this mixture would have worked out, but to run an Agile-based assignment, the resource mixture has to be set right on par with expectations.

Now after collecting all the data points, I set about to tackle the easier of the two issues, starting with the changes in the resource mix for the engagement. Although it was a very lengthy exercise with many other factors influencing the change, to make this discussion short, I have cut down the discussion to highlight only the core areas. The changes did not predominantly involve physical movement of resources although there were some cases of resource changes in the project. In these cases, the dominating factor lay around redeploying the resources to the right roles. As an example, resources with proven capabilities to progress to the next roles were moved from junior developer to developer roles, from developer to senior developer roles and so forth. In addition, there were cases of vice versa deployment. Though there were some deviations, the mix demonstrated was fit to run an Agile-based engagement. For cases that demanded physical movement of resources, this was easily achievable. Since we had been executing this engagement for long, we had a large set of resources working for different projects with the same customer. I had to counter deploy the resources within the projects such that the resource mix demonstrated in Figure 3.5 was achieved. Finally, we were able to reach the conclusion with the optimal resource mix that would be required for an Agile engagement model.

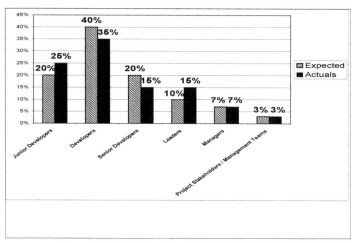

Fig. 3.5: Proposed resource mix in the engagement

The next step was to achieve an optimal resource trend in the engagement where I had to tackle issues with respect to technical capabilities, years of experience, motivation, interpersonal skills, and customer interaction skills. However, lack of some of these features in individuals has not led me to change the composition of the teams by just replacing resources. According to me, the wisest managers are those who do not waste their time trying to change people dramatically. Rather than focus on weaknesses, I believe in building on the core personal strengths of my team members and helping them to become more of who they already are. That is what I had enforced in the teams and there lay the difference in my model. I started acquainting my teams with the principles of Agile and getting them to know the newer opportunities of working in a different space, building their motivation towards the new model. In addition, individuals who had the zeal to do more and prove their skills but who had not yet been provided an opportunity were the most sought after members of the team for me. One aspect to note is that it clearly is not realistic to expect someone to become adept at all the skills overnight. However, it would be reasonable to expect a person to pick up these skills over time under the proper supervision of the right leader. Those who had the ardor for that were those selected by me for this engagement.

As a classic example, Gopal was one of the most senior and brilliant developers, a master artisan who loved coaching the teams, as well as a profound developer with heart. He was also a great learner, having the inclination to learn all the latest technologies and codes. However, during the course of this ongoing project, he was given the responsibilities of a project leader, which was not in his area of interest. His dislike of the role was a serious weakness in the eyes of the world—he was abrasive with certain people in personal interactions, which led to a bad team spirit for the whole engagement.

I had selected Gopal to work on this Agile project even though he was not recommended by the other teams. I had been warned about a situation that he had created with a client on a previous project, but I was completely positive that, given the right kind of role, Gopal would have a positive effect on the whole engagement. Wisdom should have led me to move this resource away from my team but instead, I moved ahead with my gut feeling and deployed Gopal as the senior developer for one of the modules. Although I was sure that Gopal could not change his attitude, I was sure that making him work in his area of interest—as a technical developer— would provide him the opportunity to prove his skills and he would likely excel due to his numerous technical and analytical strengths. However, for all client and team interactions, I insisted that Gopal and another team member, Sharma, work as a pair. Sharma was a project leader and had fantastic people interaction skills. In short, I did not insist that Gopal significantly improve his weakness but simply worked around them and build on his greater strengths.

However, I did make a few changes to the teams in which I moved some of the older teams out from the project due to their lack of some specific skills after the realization that they could not be successful in this kind of engagement. I had new resources join the project to create a winning composition.

In deploying newer resources into the team, I checked out a few characteristics that I had determined to be essential in the individuals. Some of these are mentioned below. This is only a partial list of qualities, technical and non-technical skills, because every individual is different:

- Strong technical skills in the area of this engagement

- Strong people management skills (customer and team management skills)

- The right spirit of collaboration and teamwork

- The ability to take initiative and the zeal to do more

- Enjoyment of work in an intense and iterative milieu

- Ability to develop and maintain worthy relationships across the team

What I was also trying to achieve with this change was to get my bulge ratio in proper shape. Although I was not 100 percent successful in arriving at the optimal bulge ratio, I was able finally to achieve a working ratio of the resource mix for this engagement. Figure 3.6 demonstrates the resource trend that was implemented after these changes.

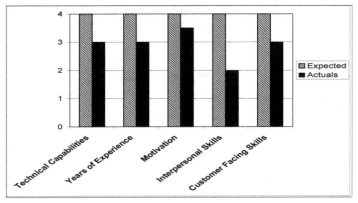

Fig 3.6: Proposed resource trend for the engagement

You have to note one important point. The resource mix and the resource trend though mentioned separately in my discussion had to go hand-in-hand in order to tackle the issues with respect to resources. Neither can be handled exclusively.

Finally, after an unexpected period of almost a month to understand and change the team composition, I initiated a trial Agile engagement. Though we faced many hurdles during the initial days of the assignment, we started to see a difference after the first month. All the teams were moving in the right direction to achieve a common goal and we were all following the Agile principles. The final lesson learned from the empirical results derived out of this experience was that in Agile practice, individuals make the difference. Without having the right set of motivated resources, you can never lead an Agile project to success. For many people, this is extremely scary, but for some, it is daily life. That is why there are such polarized reactions to Agile.

I had always followed a 4–phase approach (described in Figure 3.7) when planning any new engagements. Although the approach fits well with my Agile needs, it was never specific to only Agile projects and worked the same for any engagement needs.

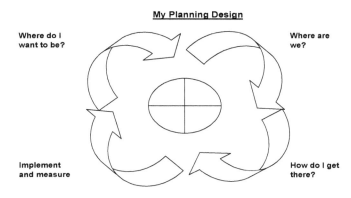

Fig. 3.7: A helpful planning design

- **Where do I want to be?** Define what needs to be achieved and what the target goals are for the assignment. In plain terms, define the goal.

- **Where are we?** Analyze where we are in terms of the engagement and the practices and processes we have been following.

- **How do I get there?** Define the plan and workarounds of the steps that need to be taken to achieve the defined goal.

- **Implement and measure**. Implement the plan and measure the effectiveness of the plan.

Observe the diagram above carefully and correlate it with the discussion for this case study. In the first step, I had the requirements from my customer to implement the engagement in Agile. This answered "where do I want to be?" As my next step, I tried to understand the currently executing engagement dynamics to understand the pros and cons of the approaches being followed along with the strategies.

As the third step, I tried to define the plan for the approach to Agile for this engagement and determined what new things to implement as well as what changes to make to the existing assignment. Finally, I implemented the plan and measured it after a predefined time.

CHAPTER FOUR

Reduce Documentation and Create User Stories

Ten years ago, if anyone would have asked me about
software development techniques, I would have told
them about the Waterfall method, the object-oriented
model, or the Rational Unified Process (RUP). I would have
even explained the advantages and disadvantages of these
models.

Today as well, in this significant phase of the emerging
IT industry, you might get the same kind of answer if you ask
the same question of a software professional. However, few
of them know that new and emerging development models/
techniques have changed the world and are now picking up
momentum and widely appreciated by customers worldwide.

The true goal of today's software development
efforts is to create working software, not documents. You
can create a very good requirement document that will be
highly appreciated by the customer, but in the end, if the

implemented software is not up to the customer's expectations, the documentation is of no use. I am not saying that we should not create documentation for the project, but in the Agile world, the major concentration should be to create working software rather than lengthy documentation. In addition, when we try to create lengthy documentation, there are high chances that the crux of the requirement is lost in the details.

From my experiences, I have learned not to force the team to create lengthy and comprehensive documents during project execution.

Documentation from the Agile methodology perspective has its place; written properly and concisely, it is a valuable guide for people's understanding. Often, it is a common practice to develop requirement specifications, system specifications, and high– and low–level designs before the start of the development phase. This process can drag on for several weeks or sometimes even months. These documents can become so huge that during the course of the project, the essence of the documents is lost and the documents become outdated when the actual development gets underway. They become cumbersome to maintain and soon changes are not reflected.

In Agile terminology, requirements are called *user stories*. A user story is defined as a small portion of the user requirement, which defines the behavior of part of the system from a customer's perspective. They are formatted in a few sentences with no technical jargon and are easily understandable by any reader. In simplistic terms, user stories are the response to, "Tell me what you require the system to do." Write down the heading of the story and a paragraph or two describing the user requirement. Another way to interpret the story card is that it creates a basis for conversation between the customer and the development teams. There should be some details in the story, but not so many that anyone who reads the story cannot clearly understand the general underlying requirements. The conversation takes place in Iteration Planning, during which the analysts and the developers ask questions of the customer in order to flesh out the details of the story and then brainstorm the tasks required to implement it.

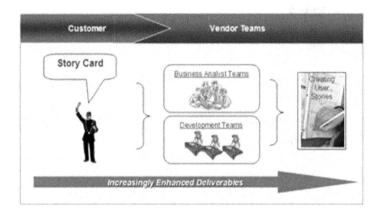

The customer elaborates the user stories and the team creates short stories of a few sentences describing the requirements more fully. What I personally feel is that the core development teams along with the business analysts should be involved in the requirement gathering sessions. This will help the technology teams to understand the requirements more clearly, which in turn reduces the cycle of business clarifications. Prototyping, especially paper prototyping, is a better technique than specifications to capture customer requirements quickly and accurately. Developers do not write user stories but do have a responsibility to review them.

There has to be at least one story card written for each function of the system and it is mandated that the business analysts write the stories in collaboration with the customer. The development teams should never be given the responsibility of writing the story cards, but they need to have conversations with the users that are then attached to the stories together with pointers to supporting documentation.

Examples of User Stories

Example 1

- Students can purchase train passes
- Train passes can be obtained from any MRT counter or can be obtained by registering at the mentioned site
- Train passes can be obtained via credit card
- Passes can be paid via PayPal™
- After you have received the pass, you can use it for one month

Example 2

- A training request can be raised by searching for your relevant course at the site
- Once you have raised a training request, it will go to the supervisor for approval
- After your supervisor approves the training request, your training schedule will be finalized by the training division and they will send you mail on the schedule of the training
- After the training is completed, you will have to take an online test which will assess you on the received training
- Your score will be sent to you and your manager
- Any training development needs will be highlighted

Each of the statements above represents a single user story. Story cards have proven to be a great tool for me during project execution. When I try to put all the story cards together and discuss them with the customer, the entire system flow is clearly understood. They also have proven very useful for me during development, providing a very concrete reference to how much has been completed and how much work is left.

A big difference between story cards and a requirements specification document is a focus on user needs. Details of the technology, algorithms, usage patterns, etc. are not documented in a story card. It precisely defines how the system should behave. Stories should be focused on user needs and benefits as opposed to specifying GUI layouts. I have always made my teams write the stories on 3" x 5" index cards, since index cards are small and they will automatically limit the amount of writing that can be put on them (which is a good thing). This forces the customer to focus on making the text clear and concise, while being as simple as possible.

During the initial start of the story card discussions with the customer, it is always recommended to have a high–level overview of how the complete system should behave first. The analyst teams break the complete system into manageable user stories and get a final confirmation on the logical split from all the stakeholder participants. As the hands-on experts on the functional workings of the complete system, they will be the right resources to judge the split. This should not be a lengthy exercise and should be accomplished in a day or two.

Next, you have to determine the priority of each story card at a high level and document who is the business owner for each story card, since for many such engagements, there might be numerous business owners for each business line or a set of story cards. I often use a scale to assign numbers (such as TC001, TC002, TC003, etc.) to each story card where the highest number has the highest priority to develop. Other prioritization approaches are possible —priorities of High/ Medium/Low are often used instead of numbers. Pick a strategy that works well for your team.

However, before arriving at the prioritization numbers your teams should first determine the technical and functional feasibility of developing each story card, such that the dependencies are thoroughly considered

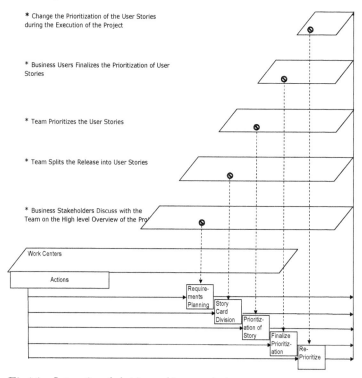

Fig 4.1   Integration of decision making at each phase

Now, the next step is to manage the network of complex relationships among the business stakeholder community. The sheer number of stakeholders can make this task of prioritization overwhelming by discussing among themselves and finalizing the priority of each story card. Since this is a distributed decision–making environment, decisions are made in a large number of forums with different business partners from each group representing the whole system. Thus, there are conflicts and inconsistencies in decisions influenced by each such business stakeholder, hence the need to integrate decision making in the program as a whole. Figure 4.1 shows a way to integrate decision making for each story card.

This is the most difficult task and it is easier said than done. Feedback and feed–forward of information and decisions at various levels are very important for the success of the project. Diverse business stakeholder participants, including the project teams, ensure that all aspects of the projects including the story card priorities are considered and agreed to in an integrated manner before starting the development. This is an important step in concurring with the schedule of the story card. There are also cases in which some requirements are absent as well. These should be given a low priority untila later pre-iteration reprioritization and planning session. Missing requirements are not specified as defects later; rather, we should consider them merely "undiscovered." Lately, Agile methods are used to refine requirements. Early requirements gathering and prioritizing is simply a place to start. It is just expected that more requirements will be added as more is known about the product. Conversely, the method of trying to gather all the requirements before starting development is not the agenda of Agile, but to gather whatever is known and document them is prompted by the methodology.

Now, as the complexity and the size of the project increase, the influence of the review system on the story card decision making increases, as shown in Figure 4.2. Review systems become vital to prioritization, performance, and schedule goals in more complex and lengthy projects where the system complexity is high and the requirements are large in number.

An effective review system on the prioritization of user stories can play a key role in the success of the project in a time bound model such as Agile and focus on problem forecasting and prevention rather than dealing with problems after they occur. To accomplish this, it is essential to integrate the decision making processes at various levels and ensure an integrated implication analysis of functional and technical issues. It is also very important to collate and follow up on the decisions made at various forums and balance the decisions between the functional and technical dependencies and the business expectations on the story card priorities.

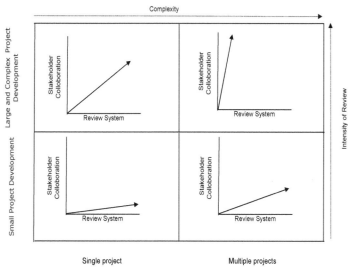

Fig 4.2   *Influences of review mechanism on projects*

After the story cards are written down and split into logical requirements, the next step is to give a high-level estimate for the stories. Because story cards contain so little information, the team will need to flesh out each story as they work with them. During the estimation effort, it is quite common to list programming tasks required to implement the user story.

Developers and the analyst(s) should collaborate in a room and work to provide a high-level estimate for the stories that are clear in nature. Group estimating has always proven successful since sharp and like-minded people work on the same piece of work. Leaders of the engagement, especially those new to implementing projects in Agile, should ensure that the estimates and changes to the estimates are tracked. I have always ordered my managers to track the estimates and the frequency of the changes to these estimates so that they would be able to apply the learning to future Agile projects.

Figure 4.3 clearly depicts how the estimates were handled and highlights any deviations to the estimates as lessons learned for all the future iterations.

Requirement ID	Requirement Description	Business Stakeholder	Initial Estimate		Revised Estimate		% Deviation	Actual effort (PD)	Final Status
			Estimate (PH)	Estimated Date	Estimate (PH)	Estimated Date			
TC 001	User Creates a Training Request	Catherine	10	10/1/2007	Not Changed	N/A	0	12	Needs Improvement
TC 002	User Approves a Training Request	Julia	8	10/1/2007	14	10/2/2007	0	14	Needs Improvement
TC 003	User is Notified of the Training schedule	Diane Frankfero	8	10/1/2007	Not Changed	N/A	0	14	Needs Improvement
TC 004	Administrator Views on the Training Needs	Bala Murthy	8	10/1/2007	Not Changed	N/A	0	8	Good Job
TC 005	Administrator View all the Requests	Cristofer Macman	4	10/1/2007	Not Changed	N/A	0	4	Good Job
TC 010	Approver Views the Training Requests	Cristofer Macman	4	10/1/2007	Not Changed	N/A	0	4	Good Job
TC 012	Add all Courses	Cristofer Macman	4	10/1/2007	Not Changed	N/A	0	4	Good Job

Fig 4.3   Estimation and prioritization of user stories

Once the stories have been estimated and the initial project target velocity set, the customer again has the luxury to make tradeoffs about what to do early and what to do late and how various proposed releases relate to concrete dates. I will not go into detail about release planning in this chapter but will discuss that in my forthcoming chapters.

Note that these estimates should always be proclaimed as "guestimates" to the customer and is not an absolute commitment. Agile methods are likely to be right about the budget involved in the current cycle, but estimating is poorly understood for the unknown requirements of the next cycle. So, how do we handle the situation of budgeting for the complete release? One thing that can be done is to negotiate with the customer to fund the project for each iteration, which is equivalent to how we fund an entire project traditionally. However, during most of such project implementations, many customers would like to know the total budget involved for a release up front. In such cases, the business users need to express all the requirements as well as they can, and the estimate needs to be adjusted by the probable cost of later changes. For instance, in one Agile project implementation, we gathered all the requirements with the best effort of the business users and the probable deviations of later changes were estimated at 40 percent. Since the cost of the project was determined as $10 million, we had to budget the complete cost of the project at $14 million with a 40 percent accommodation for later changes. Another methodology I have followed with some of my customers who were not clear on the percentage of the future changes is to gather information from similar engagements with roughly those same characteristics that I had executed and arrive at the cost of "later changes." Of course, as with all estimations, this cannot be done without considerable historical data.

During my experiences, I have not used any specific tools to capture user stories. However, if you have discovered any tools that capture user stories in a systematic fashion so that all the user stories can be linked together to form a complete system, that would be a great tool to use. In the emerging market, many vendors have created such tools that can be used for requirement capturing.

User stories serve essentially the same purpose as use cases but are not quite the same. They are also similar to usage scenarios, except that they are not limited to describing a user interface. The stories are used to create time estimates for Release Planning and replace the traditional large requirements document.

Finally, this chapter demonstrates how to design and deliver small parts of a software application in short time spans instead of creating huge documents.

In one of my experiences where I was performing a vendor management role for a major bank in Shanghai, I was implementing Agile on a couple of my projects with different vendors. We were operating in a multi-vendor offshoring (multisourcing) model in which I had my portfolios spread across different vendors and I was leading the complete engagement for my organization. I influenced the choice for the multi-vendor strategy for my organization for this engagement and attempted to de-risk the outsourcing strategies by preparing a selected list of vendors ("preferred vendors") from which we chose and sourced work. Another reason I selected the multi-vendor model was that many of my selected vendors were relatively new to the Agile implementation model and I would have the choice of dropping some of the vendors from the list if they seemed that they might be failures in the long run. The only agenda for me was to have the best of breed vendors who could digest Agile completely on my list and continue a long-lasting relationship. One point to discuss here is that vendor offshoring is a very complex strategic decision. Since it is hard and expensive to change course midstream, organizations and leaders need to spend considerable time strategizing and planning the model suitable for their specific business needs.

As part of the vendor management committee, I have to prevent myself from digging my own grave by selecting the right kind of vendor for my engagement. Since I am interested in and focused on protecting my company's interest and my own career from vendors that go bad, I had to do a thorough job of investigation to know whether the vendors had the skillsets to handle an Agile engagement and whether they had delivered similar kinds of services in the past. Even though other revenue groups did the financial stability performance evaluation, my main short-listed vendors were among those who had the expertise to implement projects under Agile and had the proven capabilities.

Finally, managing a vendor is another ball game in which I believe in driving relationships through honest feedback with a balanced perspective—the good and the bad. I had always followed a few important principles to perform good vendor management. These principles include the following:

- Defining my expectations with the vendor up front

- Using a metric–based evaluation model

- Defining a service level agreement (SLA) for vendor performance

- Maintaining an open and transparent relationship

- Participating in vendor meetings, etc.

However, I am not going to get into the details of the vendor management program now. Moving back to our discussion, the engagement was split into three major portfolios with each portfolio being executed by a different vendor. However, during the execution of one of the portfolios, I noticed issues in terms of schedule overruns for iterations. Initially, I presumed the delays were due to the complex environment in which the vendors were executing this engagement and counted on the vendor's experience to execute Agile engagements for a better result in the forthcoming iterations.

However, over a period, I sensed that there was something awry in this portfolio. I had my suspicions aroused and immediately set aside time to visit the vendor location. The agenda of this visit was to understand the methodology being followed to target this engagement and provide my expert feedback on any of the lacking concepts. Though a vendor relationship model demands careful performance monitoring and escalations, I always had believed in a good, mature relationship, whether it is with my organization's teams or the vendor's teams. I believe that being rigorous and cornering the vendors to derive a beneficial output does not help to build a positive relationship. However, I do not negate the fact that the customer has to be hard on the vendors in situations that really demand the scenario, but predominantly, the relationship should be well-placed to have a transparent work culture where the vendors share all the positives and negatives with the customer for timely and appropriate action. With this goal in mind, I always have tried to negotiate the best relationship with the vendors, trying to understand their pain and to provide them with the best possible options wherever needed.

Moving forward once I arrived at the customer location for my initial discussions, I sensed some defensiveness in the discussions with my vendor teams. I immediately gained their confidence, highlighting the fact that I was there just to salvage the situation and not to form any negative opinion about the vendor organization or its capabilities. In the end, I was really on task to find out what was going wrong in the engagement that was leading to an unsuccessful delivery, and finally, to provide my recommendation for a better output.

I set about the task of trying to understand the different aspects of the current executing model and was able to determine that the engagement had been following all the defined principles of Agile. I next set up a meeting with all the members of the executing teams (in groups) to salvage any process–related issues that they had in order to understand the team maturity on the Agile model, but I was not able to find any disconnect in this composition. I finally got in touch with the software quality assurance (SQA) teams to brainstorm a tailored process, and during the course of the discussion, one point stood out in my mind: the teams had been operating predominantly in the Waterfall model in the past, the same as many of my other service providers, and this engagement was among the first assignments to implement Agile. This, however, had not led me to judge the team as incapable of executing Agile projects as everyone has to start somewhere and has to be given an opportunity with which to start. However complex a situation might be, I always believe that hard work, process adherence, and thoughtful leadership will lead to a successful result. I next walked among the teams to understand the procedures of documentation that they had been following. During the course of this research, I immediately sensed the problem and was able to derive my conclusion of failure: I came across a glaring deficiency in the process in terms of story card documentation.

Since the teams (both management and executing teams) had been accustomed to the Waterfall approach in the past, where documentation for the requirements phase follows a systematic approach of business requirements, systems requirements, functional requirements, etc., the teams had developed a fear within themselves that creating short descriptive requirements (or user stories) might lead the engagement to failure because some of the areas would not have requirements traceability, change control, etc. For this reason, they had been creating all the elaborate documents of system and user requirements internally. This led the engagement to tend to have a longer iteration duration than planned and hence all the troubles.

Figure 4.4 depicts the planned versus actual deviation in efforts. Though you might feel that there is just a deviation of three days in the schedule, it tends to be a deviation of 14 percent, which for the completed iterations to date averaged to an overall schedule deviation of 15 to 18 percent. If you would argue with me about the fact that deviations do occur in a project and there should be a scope of absorbing the minimal level of deviations in any engagements, my answer to this is that 18 percent is not a small deviation that can be absorbed. Even if these deviations are because we have misled the engagement by departing from some of the hard and fast rules, we should correct rather than live with the problem and not compromise our goals based on the deviations.

Iteration 1 Schedule	D1	D2	D3	D4	D5	D6	D7	D8	D9	D10	D11	D12	D13	D14	D15	D16	D17	D18	D19	D20	D21
Planned																					
Actual																					

Legends		
D	Refers to Day e.g., D1 refers to Day 1, etc	
	Requirement Gathering & Analysis Phase	
	Build Phase	
	Testing Phase	
	UAT Phase	
	Bug Fix phase	

Fig 4.4 Schedule deviation in the engagement

Gaining my momentum back, I explained the most sought after principle of Agile—documentation reduction— and I set about to explain the advantages of having short user stories. I made them understand the point that everything desired by their organization's QA processes (requirements traceability, change control procedures, etc.) can definitely be achieved with the Agile model. Then, I put forth an example.

I thought it was then worthwhile to spend some time with them and have a test run with the teams for at least the next iteration, having convinced them that shorter documentation would not lead the project to failure and that whatever benefits one can get from a traditional life cycle model could be better achieved with Agile. So, once the iteration started, I mandated that the teams not create any longer documentation and concentrate on creating only short user stories which cater to the requirements. However, all the user stories were sorted out so that clear traceability of each requirement from analysis until the testing phase was available. After the end of the iteration, the teams had delivered the required software with the schedule on track for the iteration. This was not a "rocket science" strategy but simply correcting a small flaw in the process had led to the success of this iteration.

I gave my vendor PMO strict guidelines to follow these principles for all the forthcoming iterations (and I was hard here) that had been adopted for this iteration and assured them success. After a stint of two months, I saw that the project was humming along, and now I have this vendor as the most preferred vendor on my list.

The lessons we learn from these experiences are that to achieve quality output, what is needed is a quality cycle that should be built into your process from the outset, creating lengthier documentation will not help you arrive at quality output, and finally, you need to gain customer's confidence. The one main aspect of documentation is to communicate well what is happening, what has happened, or what will happen in a project. Unfortunately, comprehensive written documentation provides the lowest bandwidth of communication while placing high maintenance demands. Therefore, the goal has to be to document only what is needed as part of the process of producing working software and do not document as a parallel activity.

CHAPTER FIVE

Architecting and Designing

My experimental work with Agile—specifically in leading the team during the "architecture and design" phase and then testing them out in competitive situations—gave me the opportunity to investigate at alternatives which otherwise would have never come to light. To be allowed to compose a framework of deliberately poor architecture and design is a rare privilege. No manager would be allowed to set about such a task. The virtue of studying badly composed architecture and design is that it furnishes valuable information about what can go wrong. During my research, I have assessed the basic principles to define the cause and effect of such badly designed applications that lead to extreme failures. Before we examine the characteristics of an unsuccessful engagement with poor architecture and design, let me say a few words about the problem of carrying out this particular investigation that concluded that failure arises from unfit resources for a specific phase and badly defined processes. In addition, failure creates more embarrassment to teams who otherwise would have been successful. Finally,

I discovered that it was harder to convince the team of the difficulties than it was to convince the management.

The strain of failing can be endured, provided it does not last for too long.

Bad Decisions—a Marginal Factor

After collecting samples of consistently unsuccessful engagements from different directories and universities, I set about examining the records of the projects' performances. I was able to draw up a long list of problems, having at its head architecture and design.

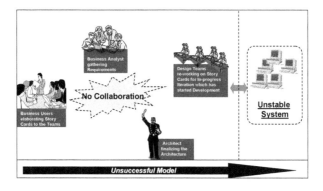

Fig 5.1: Without collaboration, projects usually fail.

The failures of bad architecture came into play in the following ways:

- A variety of people had not worked together to arrive at the solution

94

- The right kind of architect was not used

- The architecture was not decided upon at the beginning of the project and was constantly changing during the course of the project execution

- The infrastructure planners were not in place and missed the planning sessions during the initial stages of the engagement

- Programming started before the architecture was finalized

The failures due to bad design came into play in the following ways:

- Designers had not collaborated with the architect and the development teams to make critical design decisions

- The wrong kind of designers were used

- The interactive designers had not decided upon a solution at the beginning of each iteration

- Programming started before the design was finalized

Because of the sensitive nature of the subject, it is easier for me to report on the characteristics of the "bad design and architecture" in this book than it was at the time, while running the team, running seminars, or playing management games. However, at any point in time, the failure of the engagement does not lie fully with the team but evolves from the leader who has formed the team—in essence, his or her ability to judge the right kind of resources for the engagement apart from the process. The lesson is that every management team needs to have within it one person (at least) who is clever enough to form the right kind of teams and processes for collaborative teamwork. This generalization is a fair working hypothesis applicable to virtually every situation in which a management group holds major responsibility.

During my investigation, I discovered one interesting point: unsuccessful teams do not necessarily suffer from poor morale or poor teamwork. Poor morale may reveal itself—and it usually does—as a consequence of failure or of diminishing fortunes, but it should not be seen as its cause.

Ways to Successful Architecture

What I really want to elaborate upon here is that winning projects executed with Agile are characterized by inclusion of a leader who is both clever *and* creative. Creativity should be treated as an entity in itself and distinct from high intelligence and analytical ability. In this sense, creativity in a leader is more important than cleverness, but if both are combined at a high level in a single person, this is a great advantage.

When it comes to considering what can be done to overcome the list of problems leading to a failure in architecture or design, the choice should not be to deploy an ivory tower architect that hands something down from on high. Instead, the architect should understand the basics of how to design a simpler system so that it is easier to modify later:

- Architect the application on as few building blocks as possible and around easy-to-use frameworks

- Finalize the architecture before design for the first story card begins

- Changes do occur in the architecture but make it a point to alert the design and development teams of any of the architecture changes and engage them for suggestions

- Do not allow an architect to develop the framework in isolation from the design and the development teams. All the teams need to be aware of what is emerging in the architecture and should always be providing their feedback for improvement of the architecture. The principle of the alignment of authority and responsibility suggests that it is a bad idea to give a single person or a few people the power to make decisions that others have to follow without having to live with the consequences personally.

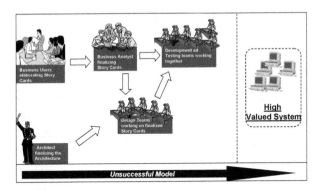

Fig 5.2: Architects need to work closely with designers and developers

I always believe in "evaluation-by-code." What I recommend is to have a proof of concept (POC) in place, built by the developers in coordination with the designers, by building a minimal amount of code around a focused area of the architecture. The goals of the POC should be to help define alternatives (while you are thinking out which way to go) and lower the technical risks in terms of the feasibility of an architectural choice. The POC should not take more than two or three days to develop for a complex scenario, and the alternatives should be well thought out for a simpler architecture.

The architect's primary job should be to develop working solutions. The best solution may not need substantial development; for example, extending packages, classes, or legacy data sources. The solution should be seen as a part of a whole that includes other systems and projects. It must be flexible enough to be changed and extended over time; however, this does mean that the architecture must be changed over time. The basic building blocks should be well defined with the aim of having an extension of the architecture, if required, rather than to re-engineer the solution. However, like other Agile team members, the architect focuses on the solution and not on documents that do not contribute to the solution.

- **Example 1**: In one of my projects, we had to re-engineer an existing application built in the "C" programming language that we inevitably had to inherit to use some of its functionalities. After a rigorous discussion among the architects, designers, and developers, the architectural decision was made to develop the project with new, more scalable and robust building blocks, keeping the existing functions as a black box but having the inherited features enabled in the new architecture. A POC demonstrated that our decision for the new architecture helped us to advance in choosing the architectural direction, reducing the risk significantly. The input from the development teams had been extremely helpful in validating the feasibility of the architecture and arriving at a concrete solution. The architecture never changed from then until the end of the project since all of the ideas of the executing teams had been well thought out and considered before arriving at the decision.

- **Example 2**: We had to compare two possible implementations of an existing multi-threaded architecture to find out which would be the most suitable fit for the implementation. What we had also thought as another alternative was to combine the merits of both architectures by combining the flexibility of the first with the utilization virtue of the other to arrive at a new architecture. We performed a step-by-step analysis of both architectures to arrive at the feasible solution considering all the complexities of both and arrived at a final solution that combined the merits of both of these architectures and inevitably discounted all the design compromises that had been in the existing architectures. We then developed a POC after finalizing the architecture. Finally, allowing for all of its pros and cons, the final architecture evolved without design complications or implementation complexities.

The proof of concept has been a very successful tool for me during any of my project implementations. It has not only helped uncover alternatives and new ideas during complex scenarios to lower the risk of changes at a later point during the project execution, but it has also helped my teams to arrive at a simpler solution.

Another key aspect for an architect to consider in an Agile development model is partitioning systems. Partitioning is not an up-front, once-and-for-all task, though. Rather than the terminology of divide and conquer, you should emphasize conquer and divide in your Agile engagement. First, a small team should prototype a small system and then they should find the natural fracture lines. They then should divide the system into relatively smaller and independent parts for expansion to the overall system. The architect helps choose the most appropriate fracture lines and then follows them to form the whole system, keeping the big picture in mind. The individual groups focus on the smaller sections.

Ways to a Successful Design

The designer on an Agile team chooses the overall metaphor for the system and evaluates usage of the deployed system to find opportunities for improvements.

Agile processes, as discussed earlier, follow a different model. They embrace change, allowing changes to occur throughout the project. Changes should be controlled, but the attitude should be to absorb changes wherever possible. Partly, this is in reaction to inherited instability of requirements in many projects; partly, it is to better support active business environments by availing them of change at any point in time. In order to make this work in design, you should carry within yourself a different attitude towards the changes. Instead of considering design as a phase, which should mostly be completed before development, you should treat design as an ongoing process that tends to sync up with the iterations of the development. This is the contrast between planned and evolutionary design.

Application design

Designers should figure out that there is a close correlation between the architecture and the design they are developing for a story card. Again, the same principle of alignment of authority and responsibility applies where an individual should not be allowed to make decisions that could have far-reaching impacts. All the teams should collaborate to validate the design and to provide feedback for improvements. During the design phase, the application designer should work closely with the technology teams and the customers. Then, at a certain point, when the focus changes during the detailed design phase, the application designer should collaborate with the development teams for review and feedback. It is necessary *not* to create lengthy application design documents; at the least, they should cover, in detail, what the classes are that should be used along with the functions. The sequence of flow can be derived from the story card, but my experience says to have a sequence of flow defined between the classes as a minimum. An Agile application design should

- Create good designs initially, and review and improve them regularly and continuously

- Apply design cleanup at regular intervals rather than all at once later in the process

- Alert the development teams to any of the design changes and engage them for suggestions

- Ensure freezing of the design for a story card before development of that specific story begins

Database design

The methodology used to work on the database design is in no way different from application design in Agile terminology. Many people during the initial days of their assignments have questioned me as to whether the iterative approach will work for their application, which is large and high in complexity because of the large database components. Indeed, many people told us that it was impossible—a troubling thought as I embarked on a large database–oriented project using Agile.

This is the most important aspect for any project as database design carries the same importance as architecture for a successful project implementation, as these are the building blocks for any successful project. However, at the beginning of the engagement and in parallel with the architecture phase, the DBAs should collaborate to define the high-level conceptual view of the database for the project. However evolutionary the database might be during the iterations, at a basic level, the conceptual model has to be decided upon before the start of the first iteration for application design and development. Once the conceptual model is derived, the physical model has to be derived in a phased manner for each iteration.

Evolutionary design acknowledges that people learn by trying out ideas. In programming terminology, developers learn and implement by trying out different alternatives to arrive at the final solution. The same principle should be applied for the database design as well. As a result, it is important for developers to have their own sandboxes where they can research and not have their changes affect anyone else. Many times, I have heard contradictions to this from DBAs, stating that multiple databases are too difficult to work with in practice. Though this may be true, in this model of iterative and continuous integration, it is possible to have multiple instances of the database so that the ripple effect of the changes made by one DBA does not hit the productivity of the other. Indeed this methodology has proven successful for me in a couple of my Agile experiences. Besides, many of my Agile teams have also proved that it is easy to manage multiple databases if they are planned accurately and timely execution of work tasks is their top priority.

Planning, designing, and creating the physical database is the sole responsibility of the DBA; however, developers can have their suggestions on the improvements and ways to simplify the design, wherever applicable.

Though developers can research and experiment in their own sandbox, it is important to bring the different approaches back together again frequently. The application should have a master database that should be maintained by the DBAs with exclusive permissions. Developers should begin their work with a copy of the master database in their individual workspaces, where they can manipulate them and finally get approval from the DBA to integrate them together with the master database. The final authority and responsibility of the database and the integration of the different developer pieces, however, should lie solely with the controlling DBA. As a rule of thumb, each developer should integrate once a day.

To have this model work, there should be very close coordination between the DBAs and the development teams:

- DBAs and developers should make themselves approachable and available.

- DBAs should make it a practice that anyone can just pop over for a few minutes and ask some questions.

- Managers should make sure that the DBAs and developers sit close to each other to facilitate communication.

However, this is not an easy approach as there will be many interdependencies in a database design, each developer having to work on the same set of tables as others, and having his or her own reasons to change the design. In the end, however, it is not an impossible or impractical approach if planned and coordinated properly. The DBA should validate changes to the design by coordinating with each of the developers, keeping in mind the high-level view of usage of the same design by other teams—changes which do not contradict with the requirements of the other developers.

In many cases, I have noticed people setting up barriers between the DBA and application development functions. These barriers should be demolished for an evolutionary database design to work. However, the responsibility and the ownership of the database design should be with the database designer, and developers should become active stakeholders for a better design.

- **Example 1**: Let's take an example in which one of my developers, Gopala Behara, comes to the office at 7 am and starts his development work by taking a dump of the master database. During the course of his development work, he predicts a change to the database is required to complete his feature. He immediately grabs the DBA to discuss this before going ahead and adding a column in his workspace. At this point, the DBA can raise any issues that Behara has not considered. Finally, after validating the request, the DBA informs all the developers through a group email list of this new requirement requirement for adding a column for Behara's code. Now, one of the other developers responds back with a similar kind of requirement that has cropped up for him the day before, and he has added a new column, which can be used for Behara's requirement as well. (However, this was done in coordination with the DBA.) This kind of coordination between the teams has helped the DBAs to arrive at a consistent database design.

- **Example 2**: In another case, G.S.M.R. had a requirement to add a column for a specific table.

He quickly verified that the column he wanted to add did not already exist in the database, and informed the DBA of the change. He then added the column after receiving confirmation from the DBA of the change. As he was working in his defined sandbox, the changes he performed did not affect anyone else's work. In addition, since the DBA was in on the change, there were no chances of fallback from the DBA on this change. Finally, at some point later in the day, he was sure of the change in the database and quickly synchronized with the DBA to reflect this change in the master. The DBA performed the change and informed all of the developers of the change.

Infrastructure Planning

For an infrastructure planner, it is of utmost importance to make all of the infrastructure decisions at the beginning of the project. It should not take long to get agreement to work incrementally on the infrastructure planning but the structure should be formed before the start of the engagement.

Finally, the Core Principles of Agile emerge:

- Architect a simpler system

- Design the entire system not just the surface

- Design with an evolutionary and user-driven process

- Design rapidly and iteratively rather than try to build a final design all at once

- Favour simplicity over complexity

- Collaborate

The crux of the model says that with close coordination between the teams, even applications with very complex designs and architecture requirements can cross any hurdles and move into a smooth execution flow.

What then are the implications of this chapter? In practice, a good mixed team of implementation level architects and designers and better planning provides the most consistently successful results. This is why there is much to be said for forming a group whose members are more specialized in their functions and abilities. Our winning teams are capable of achieving great things.

A major challenge to succeed in Agile projects is that the leaders live under turbulent conditions. Not only is the weather unpredictable and challenging, but it is usually bad. Challenges such as disrupted plans, collaborative teamwork, individuals' attitudes towards work, issues around customer availability, etc. have a huge impact on your project. The role of an Agile leader is to foster a positive attitude by creating a success-minded mentality within the teams, and to encourage implementation of proven Agile processes to enable the project to continually adapt to volatile circumstances. For the Agile leader, managing the project context is a six–folded role, navigating through the political landscape of people management and the landscape of processes, generating results, leading the teams in the right direction, managing the mood, and ensuring that values emerge with the business expectations.

In one of my experiences in which I was operating as the technical engagement manager for a prestigious project for a major training provider in the Asia Pacific region, we had initiated the project using all the principles of Agile. Some of the points that I ensured that my teams took into account are mentioned below:

- Our proposal suggested delivering the requirements in a phased approach in which each phase would accommodate two to three weeks' duration.

- I made the resources aware of the Agile principles and the demands that this process model has on each individual.

- I tried to deploy the best-of-breed resources on the teams.

- We had highly-motivated resources in the teams.

- The testing teams were ready to perform iterative testing.

- Expectations were set with the development team that we would be using iterative development.

I was controlling this assignment from the customer location and having a blend of onsite teams, which perform the initial handholding of the assignment with respect to the requirements gathering, application design, etc. We had set the initial parameters of the engagement in the right order. Where I had my offshore manager and the teams, we had kick-started the project. During the early stages of the assignment, we had the architecture finalized and had the first iteration started.

During the course of the first iteration, we were quite successful. The teams were working in close coordination and the daily standup meetings were demonstrating the value of improved communications. Coordination between the development and the testing teams was very effective with all the software pieces being tested at the right time with a very low defect ratio. After the completion of the development and testing for the first iteration, we set about to deploy the application for user acceptance testing (UAT) and the customer was delighted to see a bug–free release. They lauded the Agile process along with the teams' capabilities. The management, including the customer, apprised the teams to digest the principles of Agile into a short span to have the engagement move into a successful deployment.

Finally, we set about to target the next set of iterations with the same rigor, but over the period, we started dealing with issues. Slowly, conflicts cropped up within the design teams and between the design and the development teams. I saw this as leading the project into calamity. Though, I was operating from the customer location, I had been keeping a close vigil on the team and engagement dynamics. Many ambiguous areas had led to a tussle between the teams and each team had its own way to defend its propositions. Let me explain with an example:

Hack Mike says: Personally, I feel that presentation logic should be abstracted from the content logic as much as possible and add the navigation as a third independent entity. While I agree that hand-coding tiny variations is bad, I would propose using the XSL layouts so only a few unique pieces of code end up in a new file. I feel we will have to follow this approach.

Sharma says: As I see it, this would not work out. Alternatively, "contaminating" the XSL with conditionals will be the only possible option.

Another senior designer Chakradhar says: The point that is being missed is that with XSL, you have at your disposal any number of hierarchical data structures. To restrict the source data repositories to house only "content" is to overlook the numerous possibilities offered by XML and XSL where information can be put into the source trees—data that the XSL can use to go about building up the results tree.

None of these discussions helped to arrive at the design solution for the iteration. Each designer was trying to enforce his or her ideas on the others, creating a confrontation model.

On a similar note, one of the designers had drawn activity diagrams to depict one way to model the logic of the story card—a very common use of activity diagrams because they enable you to depict both the basic course of action as well as the alternate courses.

Another designer had a different approach where he had avoided the use of decision points and used the concept of partitions, also called swim lanes, to indicate who/what is performing the activity.

In another example, the developers were arguing among themselves and with the designers to use their defined naming conventions for the class names. This had turned the build phase into a nightmare since each had his or her own way of representing their naming conventions and in many cases, the same names were being used by multiple code modules. This was leading to major conflicts while trying to integrate the code.

Now, the bad mix of resources was not the only reason for such conflicts between the teams. I duly acknowledged and respected the resources deployed on the project. The reasons for the failure were poor planning and project management and, yes, the leader was the culprit.

As mentioned throughout this chapter, all decisions and guidelines, whether with the design concepts, development concepts, or testing concepts, should be well laid out in advance before the start of the project. This is because iterations are short on an Agile project and arguments or discussions over unresolved or unplanned principles during an iteration tends to exhaust the time available for the iteration, which in turn leads to increased pressure on individuals to maintain the schedule. This finally leads to bad team morale, and in the end, dooms the project to failure.

The only solution to this issue is proper planning.

- Before the start of any engagement, you need to ensure that an architectural framework is selected for the project. Though not completely finalized, the basic blocks should be defined and laid out, supported by a POC. The POC forms a base for the development teams to implement their features following the architecture.

- All design guidelines should be well laid out in advance before the start of the initial iteration and those guidelines should be defined by the design teams deployed on the project with a thorough acceptance from each member of the team. All conflicts in terms of the design approach should be resolved in advance and each individual has to arrive at a common understanding.

- All coding guidelines should be defined before the start of the development phase and the leader should ensure that the team understands and follows the defined guidelines.

- Each member of the team should be aware of what documentation should be created and is mandated for each phase of the iteration.

You might have understood by this time that we failed to define strategic guidelines for the design and development teams, which led the engagement to failure. I immediately set about to correct the engagement dynamic within the teams. My manager and the teams were called for a meeting and I worked with them to define as a top priority the guidelines for the design and the development phase. We had collected many best practices along with the defined guidelines from previously executed projects and tailored them to the needs of this project.

Though initially there were too many contradictions to overcome before one could define the guidelines, after rigorous brain storming sessions, the teams were able to come out with the first version of the guidelines. By this time, we had already been delayed in delivering the currently planned iteration. I immediately had negotiations with my customer to have the current iteration dates pushed back a week, informing them of some of the issues that currently needed attention as a top priority and assuring them that we would balance this delay with the forthcoming iterations. I was able to convince the customer on the facts of the current events and finally, we decided to start anew on the current iteration. Since the core guidelines were clearly defined with strict instructions to the teams on the unacceptability of any deviation, the teams set about to work on their respective tasks. Though the teams again had some confrontations on the approach, it was well managed by my offshore manager, and finally, we were successful in delivering the iteration without any further delays. The design and the development teams now understood the importance of defining the guidelines for a successful release along with the importance of collaborative teamwork.

During the next few iterations, we had the guidelines updated with all the learnings and the applicable principles. Now, the teams had developed the instinct that they could not work without having and using these guidelines. We had now crossed the hurdles and the teams were running the project in the right spirit.

The lesson learned from this example is to have the teams perform thorough groundwork before the start of any engagement that follows Agile. This is a similar concept with any of the other life cycle models, with the only difference being that you will fail earlier using Agile than with any of the other models if you try to ignore the process.

CHAPTER SIX

Setting up the Team Environment

The experiences described in this book have spanned an unusually long period. The experiments at my workplace began in the late 1990s and continued until shortly before the first edition of *My Theories and Experiences of Agile Implementations* was released.

No less remarkable than the length of this research period was that the book had ingredients of how an Agile program must perform for a successful implementation. The theory of Agile took a long time for me to fully understand with a lot of research and many experiences. However, once my experiences were shown to rest on a sound foundation, useful applications quickly followed. A renewed focus on implementing Agile theory created a momentum of its own. The problem in writing this book has been to know where to draw the line in terms of the presentation of a large amount of material.

In this chapter, I will discuss the core principles in setting up the team environment for a successful Agile implementation. The tests used in my experiments enabled me to look at a team along with various dimensions of personalities and to use these variables as the primary basis for deriving the work culture. Mental ability was of course an important variable. Before I begin to discuss the successful measures for an engagement in this respect, it is well to reflect on what results might reasonably have been expected. My measures included high–level reasoning ability; in addition, I have seen some discriminating measures of creativity based on personality dimensions. This brought me to a very interesting conclusion: *the team works well when we try to provide an uninterrupted work environment.*

The first glimmerings of doubt about the merits of this approach began to appear to me after close examination of the observed records and experiences. The dominating fact was that a good uninterrupted team environment that led to a better productive output.

Avoid Interrupts

- Ensure that your team works best when they get into "flow" mode.

- If they get broken out of flow, it takes about 15 minutes to get back in. The more interruptions, the harder it is to get back into the flow.

- Try instituting quiet time for the programmers. Turn off the phone and e-mail, no meetings, etc.

- If a question comes up, ask the teams using instant messaging tools—that's deemed OK, as long as it is still regarding the project and on-topic. This may not halt the flow like a random phone call will.

- Ensure that you plan meetings at the start or end of the day or prior to any scheduled break periods, such as lunch.

- Ensure that there are no out–of–work discussions engaged in during the regular work schedules.

In the third point mentioned above, I say to interrupt any team member whenever a requirement needs a bit of elaboration. An entirely free flowing communication policy where anyone can approach anyone anytime might sometimes move the team environment into a negative direction. However, in many cases, this has preserved the rhythm of each individual's work by maximizing the personal communication and reducing delays, but in some cases, it has led to a disruptive environment. For example, the free-flow policy of making phone calls any time has interrupted the work rhythm of the recipients thereby overshadowing the benefits over the run. However, in co-located teams, visual clues play a major role where the interruptive environment can be avoided as individuals can read the mood of the other and then enter into a conversation while ensuring their mate's "flow" mode is not broken down. However, for a distributed team, the scenario is more complicated. Since in a distributed environment the visual clues are missing, having an undefined communication channel would many times spoil not only the relationship between the teams but also the benefits of the Agile process that defines the procedures of continuous interaction. This situation thus demands that definitive protocol be defined for communication that would preserve the paradigm of the Agile approach.

In my experiences, this was achieved by defining a timeslot that is comfortable for both the onsite and the offshore teams where people can freely call each other on the phone or use other messaging tools.

Outside these preserved hours, it should be encouraged to have a non-disruptive work environment where individuals can concentrate on their pieces of work. When such protocols are defined for all teams, it leads to an environment where each individual organizes his time for a better productive output. As an example, such timeslots should be defined during the evening hours typically from 5:00 PM to 7:00 PM IST if the teams are spread across India and U.S. time zones. For other locations, a similar slot of convenience should be set up since, depending upon the time differences, the timeslot becomes a key enabler for communications, and the team should not spend extended hours in the office because of these calls.

On a similar note, the practice of extended working hours by the teams either onsite or onshore, should be avoided, even though it might be either to maximize their productivity or meet their defined goals. This "solution" typically would arise in situations of looming milestones, especially for code deliveries. However, people always working late-night shifts lose touch with their colocated teams. Another most important factor is that to push someone beyond a certain time limit tends to diminish their productivity as a whole. Jetlag from the previous day carries forward for the next two days and to get back up to speed again, there would need to be overtime work in the forthcoming days, which would again impact productivity, ending up in a Catch-22 situation. Always institutionalize smooth and reserved working hours for the team and ensure that the team avoids extra working hours so that it does not stretch the constraints preserving team motivation and productivity.

Once I was called to consult on a project that was mysteriously staggering. I started to understand the engagement more in detail and started to walk from person to person, interviewing them, and finding out that the composition of the team was highly experienced and motivated. After not being able to ascertain the problem, I sat to think for a few minutes and what finally struck me was the team coordination. The senior people had corner seats and the team interacted only a few minutes each day. Even though the planning was perfect and their methodology implemented all the best practices of Agile, there was a disconnect between the teams. I provided my feedback to the project that the team should sit together and have regular stand-up meetings. When I returned a week later, I sighted improvement in the engagement and the project was humming along. I took away two lessons from this experience.

No matter what the client problem is, it always has to do with people. Technical fixes alone are not enough. The other lesson I learned is that the team should be motivated to sit together to communicate the essence of all their learnings. Make your workspace about your work. An interested observer should be able to walk to each team member in the team space and get a generic idea of how the project is going in 20 seconds.

Daily Meeting

- Ensure that the teams sit together in close proximity.

- Daily stand–up meetings are necessary but should not last for more than 20–30 minutes.

- Ensure that you get to the point on what needs to be done for the coming period and not discuss future forecasted work as this will lead to unintentional loss of time.

- Developers/designers/requirement analysts explain all the updates in their respective areas:

- ▶ What they have done since the last stand–up meeting

- ▶ What they plan to do until the next stand–up

- ▶ What got in the way of doing their work

Daily meetings are defined as daily stand-up meetings in Agile terms and the word "stand-up" has relevance within itself. The team actually has to stand up to keep the meeting short.

Nevertheless, I was surprised, given the apparent simplicity of the stand-up meeting, to see that in one of my engagements these meetings were not working. It was immediately obvious to me why they were not working but the team was not aware of the failure, since they had not completely diagnosed the principles of the daily meetings of Agile. The reasons for the failure were that these meetings were focusing more on process improvements and the technology/functional discussions were making the meetings longer than an hour. In addition, most of the teams arrived at the meetings late.

Technically, in order for the daily meetings to be successful, the meeting agenda should be short and the purpose of the meeting should be to show the teams' commitment to themselves and to other members of the team. Sharing commitment by the teams is more important than sharing project status. This does not imply that status updates should *not* be the primary goal for the daily meeting but as a perception, this is secondary to team members publicly committing to each other and sharing obstacles that prevent them from meeting their commitments. The entire teams come together to bear the responsibility of any obstacle a team member faces since they all work towards a common goal.

Doing stand-ups with prolonged discussions on individual obstacles, or what we term as "issues," should be avoided. However, each individual should be given a chance to raise his issue and have it noted, but further discussions should be parked from the daily stand–up meetings. Resolution of these issues should be in a separate forum which you can call **"Emergency show-stopper meeting."** These types of meetings still should not become a regular activity.

Status update and tracking should be the primary goal for the leader and work performed should be reported as a team update in which the leader takes notes on the progress. These meetings are a differentiator for Agile. During these meetings, the leader also has a chance to call attention to the backlog items as well as plan for the day's tasks to set the team and the spirit in the right direction.

Now we will discuss how to tackle the situation in which status updates are provided to various representatives of the organization including the senior and the top management staff. Creating status reports and graphs, or organizing long calls with the management is indeed an overuse of time that I personally do not condone. However, this is a crucial aspect, as people from different communities who represent the project directly or indirectly would like to know the progress of the engagement and most importantly can provide valuable suggestions to the betterment of the project. However, communicating status in multiple meetings and creating reports for each such meeting is time consuming and a duplication of effort.

So how do we mitigate this risk in time management? The answer is quite simple. You, as the leader, should emphasize non–duplication of effort and simplicity in communication. Therefore, replace the majority of stakeholder status meetings with the daily team status meetings and get the relevant management teams to attend the status meetings at regular intervals. This not only provides a high–level overview of where the project is heading but gives them the micro–level details that show the health of the project.

The above holds true for stakeholders who directly influecne the project but those not directly involved might disrupt the whole behavior of the meeting. This emphasizes the need for another forum that would still be required and would fall outside the scope of the daily status meetings. You, as a leader, are the right person to determine the involvement of the relevant individuals in these meetings. Another point to note is that your daily status meeting agenda might be hijacked by having too many people attend the meeting. Typically, meeting participation should not exceed 12 participants.

In one of my engagements, I really got a very bad taste of Agile execution when the leader of the engagement made the iteration kickoff and the daily meetings nothing but sheer agony. I was beleaguered initially, but quickly I was able to attribute the fact to the leader's inexperience with the Agile methodology. The meetings used to be protracted and futile as the entire morning was wasted by the leader. After the first few discussions, I stepped in to explain the model to my leader and helped him to keep a healthy pace to the meetings. This helped him to straighten out the project management process and eliminate much of the wretchedness.

My manager quickly was able to understand the principles, values, and importance of many of the processes. He explained to me that after a few days he was able to derive a solution with a couple improvements, which continues to amaze me to this day.

- A leader's role is not to find a substitute for any individual who loses focus towards a specific goal but to lead his coworkers into the right direction for overwhelming output.

- People learn from mistakes and until individuals learn and improve, they should be given equal importance and should be provided with the freedom to improvise. This is the team-motivating factor.

- When applied cautiously and intelligently, a process can prove itself to be of utmost value not only to the project but also to individuals as well.

- A process gains its importance only when the team understands its importance and actively adapts and applies it in its day-to-day work.

- Everyone in the team needs to adopt the process. A leader cannot force the team to use the process but should work with them to adopt the process.

- Eliminate processes that do not move towards improvements.

A stakeholder is anyone who affects the success and failure of an engagement before, during, or after the engagement. Agile recommends having all of the stakeholders available to team members for consultation on an as-needed basis. Stakeholders from the user group must always be available to the project teams. The application users should spend maximum time on the project at least during the initial stages so that the teams understand the end objective of their assignments. Effective business user management is the heart of leadership for any Agile–based project. The primary role of an effective Agile project manager is to manage the network of complex relationships among the business user community.

- Ensure that your business users are always available for the team, for all the clarifications that they require. Analysts can also be effective go-betweens.

- Ensure that the customer elaborates the user stories (requirements in Agile terms are defined as "user stories") in a perceptible structure and extends the stories wherever possible.

- Business users should review the user stories as understood by the analysts.

- Users/customers read out stories for the iteration.

- Developers split the stories into tasks and discuss with the customer wherever required.

- Any unsigned stories should be deferred to the next iteration.

- Ensure that the customer works to define the next set of user stories one after the other. The analyst teams should have the latter set of user story ideas organized before completion of the former.

- If there are users assigned to specific modules, reassign your team members to the specific user groups.

- Ensure that the customer tests the application at predefined intervals after each iteration.

In summary, in an Agile implementation, you are dealing with a complex and confusing environment. It is easy to lose sight of where you are heading if the engagement is not monitored with a proper objective. To maintain your focus, you need to rely on the business values and the goal of the project. Some of the values that are part of my experiences are

Clarity of function:

In practice, clarity of function explains not only the goal but also why it is being done.

Results preference:

To apply the value of result preference means to focus more on completing the activity rather than tracking the activity.

Early implementation:

In practice, early implementation means providing the end users with something they can put to use as early as possible.

CHAPTER SEVEN

Extended Programming and Testing

A gile is a holistic model that is build around an integrated set of principles, values, and practices that accelerate performance on all three levels—individual, team, and organization. It includes the methodologies, principles, and techniques necessary to reap the benefits but unless these are firmly rooted within each individual, they are about as useful as tumbleweeds blowing across a barren prairie. Do not fall into the tool or methodology trap. This is not a fill-in-the-blanks project execution. As a member of the Agile team, emotional intelligence is one of your most important assets. One way to understand the parameter of emotional intelligence is to perceive emotional ignorance. Emotional intelligence is the ability to be sensitive to others' feelings and respond in the right fashion based on the situation in which you find yourself.

I used to think of myself as a master in emotional intelligence until my wife proved me wrong. It was one day during the past year when I had overcommitted and was trying to deliver multiple simultaneous projects. I was working in the capacity of delivery manager and I had just been married for a few months. One evening (or rather around 12:30 am) after a heavy workload, I came home and knocked on the door. There was my wife, Vinni, waiting to greet me. She threw the door open, looked at her watch, and lit into me: "Do you know what it is and where were you till now? Don't you understand that someone is waiting for you at home? Why is this getting into a regular habit for you? I missed my favorite movie because of you." So, in true *Men are from Mars, Women are from Venus* form, I calmly opened my laptop and showed her all the work I had been doing until then and how the different issues in the deliveries were hitting my project teams and me, preventing us from having a smooth delivery. I even showed her a couple of escalations from my customers just to make the fact clear to her that I was struggling hard to get things in place at work.

However positively and truthfully I approached it, she came back to me with the same response. She expected the same commitment with her that I had at work. I immediately realized that my approach of being up front with the facts to her was right, but the timing was wrong. In a similar situation, after a few days, I applied my learning. When the door opened, without her saying a word, I immediately put my hand on my head, pretending to be very wired and sorrowful for my delay in not getting home on time. After a few minutes, I apologized for the delay and acknowledged her feelings. I said "I would rather resign my job instead of worrying you," and this put her in a positive mood, not bothering about my delay, but consoling me.

Like people, projects can be in a good mood or a bad mood, and if you were able to draw the fine line of applying your emotional intelligence at the right time, you and your projects would end up in a seamlessly integrated and successful spirit. In the end, a project is comprised of individuals with unique needs.

Now, before going into the details of how to target the programming and testing approaches for an Agile–based engagement, it is important for me to define the team composition in this aspect. The development and testing teams play a major role in the success of any assignment, not merely because these teams tend be high in numbers but also because their contributions form the core output of the project.

Development teams should commonly be made up of members holding particular interests. They are there by virtue of the responsibilities and work they have represented in their experiences. No overall sense of design governs the composition of the group that, in human terms, is little more than a random collection of technology resources with as wide a spread of human foibles and personal characteristics as one might expect to find in the population at large. Nevertheless, what I have established, or endeavored to demonstrate in this book, is that the compatibility of members of the team is crucial to its effectiveness. It is no less important that the members of the team be technically competent than to have a high level of motivation to work in a time–sensitive environment that allows individual contributions to impact the success of the engagement.

My experiences and research have given me leads on how the subject of compatibility within the teams should be approached. Methods and techniques might vary from time to time but the underlying concepts are the same throughout. To guide us in the design of our teams, especially development teams, the principles mentioned below should be considered:

- Members of a team can contribute in two ways to the achievement of a set objective: they can perform well in a functional role by drawing on their professional and technical knowledge as the situation demands; and they have a potentially valuable team role to perform, describing a pattern of behavioral characteristics of the way in which one team member interacts with the other in facilitating the progress of the team.

- Each team needs an optimal balance, both in terms of team roles and functional roles.

- The effectiveness of the team will be accentuated by the extent to which the members correctly recognize and adjust their behaviours to the relative strengths of their co-workers in terms of both skills and the ability to engage in team roles.

- A team can deploy its technical resources to its best advantages only if it has a good team spirit.

- Good personality qualities along with good skills will extend the likelihood that team members will succeed with others, ultimately leading to success in their assigned tasks.

What I want to derive from the above discussion is that the degree of success for an Agile–based engagement depends on like–minded people working together to achieve a defined goal. Since the percentage of resources deployed for the development and testing phases for most projects are more than for any other phase, having like personalities will lead to a smooth execution of the development phase. Composing a team is a delicate balancing operation.

I widely promote the use of smaller teams for an Agile implementation. A common mistake is misunderstanding this to mean that you should only implement Agile on smaller engagements, which is not the essence of my words. What I intend to say that you should split the team logically into smaller teams and assign a leader to each team so that the focus on the assigned tasks is not missing. I have implemented the Agile processes and methodologies for very large scale engagements of $20 million, but the logical split on the composition of the teams for each task was never more than eight to ten members with an internal leader for each team. I have used this structure for both development and testing phases.

Now, drilling into the details of the programming phase, here are few of the principles to be followed:

- Pair Programming, which means two people working on the same piece of code so that one of the pair will be focused on writing software while the second is conducting a concurrent code review to look for defects early in the development cycle. One person is the driver; the other is the navigator. The navigator should be able to find alternatives, check for conformance to standards, and prevent the driver from writing unreadable and unreachable code.

    I remember a classic example in which the leader of an Agile engagement who claimed to understand the methodology to the fullest experimented with the pair programming by deploying two resources on a single piece of code. At the end of a whole week, he found out that this piece of his experiment was not working positively and he was wandering around cursing the principles. At that time, I was operating as the engagement manager for this assignment so I dropped in for a stand-up meeting. Within no time, I was able to understand the reason for the process failure. The leader had assigned a pair in which one was an expert programmer and other a slow programmer. Are you now able to draw your judgment of what went wrong? Let me explain.

Since the pair was contradictory to each other in terms of speed, the results were just as negative because the fast programmer had to wait for the slow programmer to complete his code. This led the senior programmer to frustration, as he had to return to his own programs, leaving the junior developer to complete his work. This also led to frustration on the part of the junior developer where he was trying to match the speed of the other but was having issues because of his desire to contribute more to the project. Of course, it was not the fault of the slow programmer since he was a junior coder, paired up with a senior developer full time in the pair programming concept.

Pair programming does not necessarily mean having two people sit together on a single machine to create the work product. Where I have experienced a better result is to make two developers responsible for the same code and one is the reviewer for the other who is creating the work product. The reviewer and developer would reverse roles on another work product, balancing the equation. Another variation on pair programming, which will definitely be enjoyed by senior developers, they should be mentors to junior resources. Most senior staff will enjoy this role as long as they can return to their own programming world for sufficient amounts of time to make up the shortfall in work or productivity. Developers choose a task (part of a user story); and the leaders choose their development partners for the day. In some cases, the developers themselves should be given the privilege of choosing their partners – this should be determined by the leader on a case-by-case basis.

- Ensure that the coordinators work on overlapping shifts with the teams so that handover/takeover of the development happens in the regular stand–up meetings and that the coordinator is always present during the handover sessions.

- Compiling and deploying code everyday/every week are irksome tasks for the developers. Reduce as much of the manual process as possible and deploy all the sophisticated tools available to reduce boredom and the careless error that it introduces. Ensure that code is deployed into the integration environment at regular intervals, which will avoid later integration issues.

   If you have the luxury of having a dedicated integrator, my suggestion would be to deploy such a resource. This will allow the developers to concentrate on their areas of work. All guiding principles for an integrated environment should be provided to the developers so that the code developed by the developers can be integrated easily.

Five Golden Rules of Coding

- Ensure customer availability during the development phase. Have the developers discuss with the business analyst and the customers periodically during the development phase.

- Always follow pair programming.

- Code unit tests first.

- Follow coding standards and update at regular intervals.

- Release code to an integration server once it has been unit tested and test it again on the integration server—only then is your code said to have completed the cycle.

The Testing Strategy in an Agile development approach is very different from a traditional bureaucratic methodology. I might have developed a very robust system but if testing lags, its focus in all efforts leads to a disaster. How Agile testing differs from a traditional approach is by performing iterative testing in line with the iterative programming for each iteration with no lag in time.

In smaller engagements, the business analysts and the developers themselves form the testing teams, but for a high-density project, the need for separate and dedicated testing teams plays a very crucial role. In one of my smaller engagements, my business analysts would do a few rounds of testing each time the software version was released. Business analysts understand the business requirements of the software and test it to verify that it meets the requirements.

However, the type of project you are executing decides the differentiation of the team composition.

If you hear the delivery teams talking to you about the tests in Agile projects, the first question should be to ask whether the application focuses more on the business tests or the technology tests. There are some essential differentials between business-centric tests and technology-centric tests and each draws a separate line of team requirements. In business centric testing, you will be more concentrated on the business logic flow. If you are having a discussion with the business user on the progress of the testing, you will be discussing in essence how a specific test was successful or failed. For example, "All the financial advisors' names along with their contact details were displayed after I clicked on the 'Get all advisors' link, but I had added ten advisors, and in this page, I am able to see only five advisors!!!! An error."

A technology-centric test will focus more on the domain of the programmers. A sample in which your application should work well on a distributed database will be one set of compatibility testing you have to perform. Another example are tests in which your screens should exhibit the same behavior in both Internet Explorer and Firefox to ensure browser compatibility.

For any engagement, both types of tests will play a predominant role but the outputs to these questions will be the judging factors to determine the right mix of business and technology testers on the testing team. A huge domain-centric financial application will require the dominant population of testers to be the functional experts.

Testing activities vary throughout the lifecycle for an Agile–based development model. During the early stages of the construction phase, a thorough analysis has to be performed to identify the right set of resources, tools, and the framework for the engagement. It should start with the installation of all necessary tools and software, aligning the processes, and getting the testing labs in place wherever required. An interesting point to be noted here for better resource planning is that you will discover testing during construction iterations will take considerable time and will enable you to do less testing during the end game. Hence, the resource ramp-up and ramp-down should be planned with these things in mind.

During the end game, you will have to perform the final level of testing on the application, including full system and acceptance testing along with thorough regression testing. This is true NOT only if you are legislated to do so or if your organization has defined service-level agreements with

customers who require it. A final test on the release will give
a high degree of confidence in the engagement and will very
likely result in success. However, if you have tested effectively
during the construction iterations, your final testing efforts will
prove to be straightforward and quick. If you are counting on
doing serious testing only during the end game, then you are
likely to end up with a very bad release because your team will
not have sufficient time to act on any defects that you do find.
Remember: process–driven Agilists test often, test early, and
usually test first. Given the fact that in an Agile development
approach, relatively short turnaround times are the goals, it
is extremely important that the team be clear on what needs
to be tested. Even though close interaction and innovation
are advocated, clear directions and goals on the tests should
be communicated. While each team may have its own group
dynamics, each piece of code should be unit tested by the
developer before passing it on to the testing teams. Since the
development is iterative, the next release of the code would
potentially have modifications to the previous one. Hence, it is
extremely important to run your regression test suites for each
completed iteration.

Test automation also gains importance due to
short delivery timelines. The most important factor in test
automation in the Agile development approach is to at least
have the system automated for regression testing, which will
reduce the boredom of the testers in performing the same
tests over and over again for the previous iterations.

- Ensure that the test teams are available as required
  for the development teams so that the developers
  can push their code to be tested.

- Institutionalize a process in place so that the
  tester who is performing his tests on a specific
  piece of functionality knows who the owner of
  the code is and can pass his testing comments to

the respective developer or request clarifications. This avoids the lead time of communication gaps. However, the defined team hierarchy should not be compromised and it should follow due process.

- Deploy all the required test management tools. Since the iteration durations are small for Agile methodologies, one needs to repetitively run tests. It is much less expensive to fix bugs early in the cycle rather than later.

Many of today's software-based projects and products suffer from usability challenges after the final implementation. Usability is a quality attribute of a system that mainly encompasses end-user satisfaction. Your system may deliver a very high performance output but unless the usability of the application is acceptable by the customer, all the efforts are in vain. However, with Agile development being highly demanding in terms of shorter durations for the iterations, complete principles of usability testing would not prove successful without watching people try to use something for its intended purpose. However, these can be experimented with during the "Pre-user acceptance testing" phase. (I will be discussing the "Pre-user acceptance testing" phase in detail below).

The approach I have been using to solve this problem is to design a prototype before the programming effort. I have experimented with and proved successful the use of paper prototypes for usability testing. Testing based on paper prototypes and early versions of software was added after my first release of an assignment in Agile and was carried forward as one of my best practices, resulting in a significant reduction of usability-related rework. What my testing teams used to do is create the paper prototype and get approval from the customer on the look and feel. The paper prototype became a tangible representation of the project vision and was used in many ways that contributed to the resounding success of the project.

Often, user acceptance testing is the final stage of testing before product release or implementation. However, for an Agile development paradigm, user acceptance testing (UAT) runs in two passes. One is called "pre-user acceptance testing" and the other is called "final user acceptance testing." Each phase has relevance by itself, presents specific challenges for testing professionals, and requires a different approach.

Even though the testing teams have 100 percent coverage on the functionality for a specific iteration, the phase is not deemed as complete. "Pre-user acceptance testing" by the business user evidences that the functionality of a specific iteration is acceptable. Ensure that after the test teams complete their testing and bugs have been fixed by the development teams, that business users perform their job of testing for each iteration and provide feedback.

- Keep the system running while also producing new iterations.

- Ensure that business users are performing complete testing of the released iteration and pass on their comments to developers.

In "final user acceptance testing," the business users perform their refined level of testing (sometimes called rigorous testing) on the complete application at the end of the release to confirm the acceptability of the complete application as a whole. This generally does not lead to an extended test/ fix period, as the business users would have already tested the application during the iterations.

Finally, what it boils down to, is that every stakeholder in an Agile–based engagement is equally responsible for the success or failure of the project. Any delay or inefficiency in performing tasks by any individual representing the project, whether it is someone on a development team, a test team, a business leader or the technology leader, will contribute to making the whole engagement fail. A people focus is more important in an Agile–based engagement than a process focus.

I recall one of my experiences where I was called in to consult on a project that was operating under the Agile process model and which had been encountering continuous failures in achieving the desired positive yields. The senior management teams of the engagement got to me with their words: *"We understand and implement all the principles of Agile without compromising any. Initially, for the first iteration, our failure was attributed to the fact that the teams were new to the Agile-executing model and we would improve over the next set of iterations. Now we have delivered the third iteration, almost a month has passed, and we are seeing the same trend towards failure. Agile will probably not fit this engagement."*

I explained my ground rules to them and convinced them that together we would try to make the engagement successful. If some of the principles of Agile did not suit this engagement, we would try to tailor the processes and the methodologies that had been followed. However, I had been wondering why this assignment had been failing, since I had been aware of this engagement from its inception and I considered this as one of the best engagements on which to implement Agile. However, I started to understand the concepts the team had been following to try to adapt this engagement to the Agile process model. After two days of continuous discussions with the managers and the executing teams, I reached a few conclusions:

- The executing teams were well suited for the expectations of an Agile project. They had more than the minimum required level of understanding on how an Agile project should be executed. They were highly motivated and the majority of the staff were technically competent with good management and interpersonal skills.

- The leaders of the engagement had complete comprehension of how an Agile engagement should be handled with a thorough understanding of the principles of building team spirit.

- The project delivery model had been well thought out and the team had been following an iterative approach with each iteration spanning two to three weeks.

- Customer availability was never an issue as the users provided their complete support to the teams around the clock.

- The development and test teams were all well placed.

I still had some discomfort, however, and my instinct was telling me to better understand the procedure we had been following before applying the iterative schedule for the engagement. I called a meeting with the project managers to understand the approach they had been following for the iterative development of the project. They presented their estimating and scheduling sheets to me with the details of how they had arrived at a logical separation of all the requirements, and how they used requirements to derive a schedule of two to three weeks for each iteration. The approach was well thought out and they had all the critical and independent business functions delivered during the initial iterations before moving to the next set of requirements. What caught my eye during these discussions was that though the iterations were well planned, there had been a small error in understanding the concept of iterative development and testing. This error lay in the planning for functional testing at the end of each iteration. Figure 7.1 demonstrates the iteration model they had been executing.

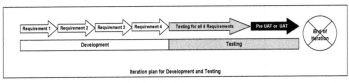

Fig 7.1 Current iteration plan of execution

Without reading this section further, are you able to understand now what the issue had been in this current model? Let me explain.

The model these teams had been following was the approach known as "code first and test later," which is not the approach for an iterative development model or specifically for an Agile–based approach. The principle the teams had been following was to complete the development of all the functions for a specific iteration and then distribute it to the testing teams for testing. Refer to Fig. 7.2 for a snapshot of the schedule defined by the team.

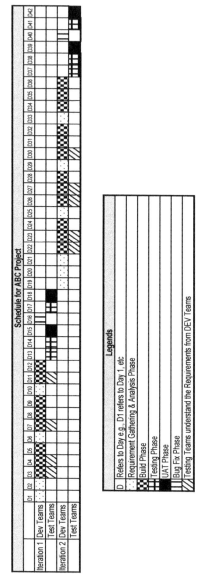

Fig 7.2 Schedule defined for the project

140

The schedule clearly contradicts what Agile proposes. Testing was initiated only after the development was completed for all the requirements in the iteration, which was one of the most important flaws in this model. Due to this flaw, a ripple effect was created in two cases.

In one case, the testing teams were pelted with the complete release in a single shot at the end of the iteration. This led to huge stress on the teams and cases were reported where some of the functions were not tested to the active satisfaction of the test teams. This also led to UAT defects that were reported by the customer. These would not have cropped up if the test teams had been given enough lead time for testing.

In the second case, the bugs reported by the testing teams had been accumulated at the end of the release (or more precisely—the end of the iteration) for the developer to fix, and finally, after many iterations, there were a huge number of bugs that had been reported by the testing teams. This had led to schedule slippages where the developers had been working extended hours to fix the bugs. As the duration of the iteration is small, apparently the bug fix slot is small. Another area of concern that the customer had escalated was regarding the high number of defects. Incidentally, this was again due to the same factor of testing at the end of the iteration. This is due to the bug fix window scheduled at the end of the iteration. In cases where the number of defects used to be large and correspondingly the time to fix was less, the developer tended to to inadvertently introduce new defects while trying to fix the existing ones.

Another important point you should have understood by this time if you have ardently observed the schedule derived in Figure 7.2 for this engagement is that the iterations follow a sequential model in which the first iteration follows the next in a phased manner. This also had a negative effect on the time boxing where the duration of the iterations had been extended to an unnecessarily long duration which otherwise would have let the project be completed much earlier than what was forecasted. Another result of these time lags was to leave some of the development resources without any assigned tasks for the specified duration of the bug fix phase, specifically pointing to functional owners who had no bugs reported from the testing teams.

Finally, after a thorough review of all the other processes and confirming my satisfaction over those, I confirmed my conclusion that fallacious iteration planning and incorrect scheduling were the two single points of failure for this engagement.

Now I set about providing my recommendations to the team. I had taken a few examples and made the managers aware of some of the important principles in an Agile iterative development phase. Testing the application at regular intervals is crucial for a successful Agile implementation. Figure 7.3 demonstrates the model of the iteration planning that I had recommended for this assignment.

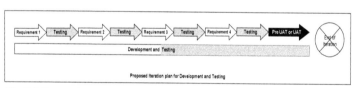

Fig 7.3 Proposed iteration plan of execution

This model also provides the testing teams enough lead time to have a thorough testing of the application throughout the iteration so that they are not bombarded with the complete release all at once. This lets the team have focused testing time without undue pressure and the output of this model is higher quality and productive testing.

Next, I focused on the scheduling aspect where we were left with some gaps between the iterations that had left some of the developers with idle time. This was fixed with a stringent schedule where iterations coincide with each other. Figure 7.4 demonstrates the sample schedule that I derived to make the manager aware of how the iteration scheduling has to be planned.

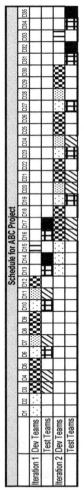

Fig 7.4 Proposed schedule defined

If you compare the important distinctions between the models with the existing schedule and proposed schedule, there are three major differences:

1. I had introduced the concept of testing the application at predefined intervals once a requirement for the iteration had been coded and unit tested by each developer.

    Advantages: The advantages the engagement had with this approach were that the testing teams were not bombarded with the complete release all at once, and the developer did not have to wait until the end of the iteration for his functionality to be tested and bugs reported. This helps each developer to balance his work, providing him the lead time to fix the bug and re-release the functionality for re-testing along with working on his allocated tasks. In addition, since the leader can anticipate the pattern of how much time each development resource is spending on the bug fixes, he can perform a root cause analysis (wherever required) on the bugs and formulate a plan to avoid recurring errors. This also gives the leader a chance to re-plan his resource loading during an iteration in which the developers are lagging behind on their assigned build task due to bug fixes. They can be provided with opportunities to share their workload with the other development resources who are relatively less loaded.

2. I introduced a phase called "bug fix or build" which demonstrated that if there were any bugs that had been reported by the testing teams, the respective developers fixed those bugs

immediately or else continued with their assigned build tasks for their next set of requirements.

Advantages: The advantage this approach had was that it demanded that the leader plan the iterations carefully, accounting for the fact that each requirement that was developed and unit tested by the developer needed to anticipate some errors in testing and that there should be enough planning to accommodate time for these bug fixes apart from the regular development work. This points out the fact that there should be a scheduled window after each build where developers can perform bug fixes. For builds that have passed the testing phase with zero bugs, the respective developers can concentrate on working on the next set of requirements.

3. Scheduling of iterations had coincided where the next iterations started before the finish of the former.

   Advantages: The advantage to this approach is that after each developer has completed his or her development work, he or she is assigned to the next iteration immediately ensuring that there is no loss in productivity by giving idle time to the resources. However, any bugs reported by the testing teams or bugs reported in UAT will be handled by the respective owners from the developer community but will leave the rest of the teams to concentrate on their activities for the next iteration. If you compare Figure 7.2 and Figure 7.4, you will note an important point: there is a productivity gain of seven working days with the latter model. You can also build your schedule so that there is built-in time for the developers to have some down time before the start of the next iteration, so that they initiate the next iteration with a fresh mind. I leave this to the individual managers as to how they want to effectively plan their time and, at the same time, have the team motivated.

One important point in this case study that I thought I would highlight is the phase marked with diagonal lines. This phase has a very significant value in the iterative model. What it entails is the testing team's continuous collaboration with the development teams. During the development cycle, the testing teams should make a point to approach the development community at regular intervals to understand the requirements and prepare their "test approach." What "test approach" means is that the testing teams understand the requirements and prepare themselves with the cases to test the functionality that they might receive in the next few hours or days.

What was inferred out of this experience is that there is no single principle in Agile that can be compromised. Each principle itself may lead the project to success or failure. You can clearly observe in the above case study that the team had implemented all the best practices to follow Agile but a small deviation to the iteration planning had led the engagement to initial failures which, if not rectified at the appropriate time, would have made the project into complete disaster.

In the case of Agile, the need for method adaptation is to be made explicit and this is the strongest principle each leader has to follow. One of the fundamental ideas of Agile is that, although all the processes cannot be implemented in an engagement in a cookie-cutter fashion, you should intuitively see what tailoring to the principles and guidelines will best fit your model. You can tailor a process but never ignore it. I do not have an experimental report in which I have followed *all* the textbook principles of Agile to be a successful Agile leader. What has led to my success is a thorough understanding of the Agile processes and tailoring and adapting to these based on the needs of each engagement.

Agile development can be a very exciting and invigorating approach, if all the key aspects of execution are followed. Some projects suit Agile development more than others. It only depends on your intelligence as to how you approach an engagement. Collaboration and visibility can provide a much richer experience for the teams who set out to develop great software products following Agile. Agile development can be a lot more enjoyable than any other method that involves a lot of overhead (management overhead, documentation overhead, etc.) — those that are less flexible by nature, and those that provide the least opportunities to correct mistakes during the course of the project. In Agile, you can drastically change the model with the experience learned in the first iteration, and that is what Agile is for. Moreover, when people enjoy their work, it is amazing what they can achieve!

CHAPTER EIGHT

Agile Quality Process

In today's tough competitive market, how do you win your share of business opportunities? How do you outperform competitors who are as competent as you are and who essentially sell the same kind of services as you do to your customers? How do you differentiate your organization and your services when your rivals imitate your ideas almost as fast as you can innovate? How, in the toughest global service market the world has ever seen, do you maintain and sustain a distinct value to keep yourself afloat in the competitive market?

These are some of the questions I was asked in a management seminar conducted in Detroit on the "Road to Success for the Trends of IT." For the past several years, we all have worked towards winning customer confidence by developing the best products and services. However, in today's competitive market, almost every organization is capable of providing the same kind of services to keep customer expectations on par. So what is the differentiator we can have within the organization to be ahead of our competitors?

The answer to this for me and my organization was to follow some of the methodologies that are considered to be industry leading and those that most competitors have not yet put into practice. One such methodology with which I had been successful and considered a differentiator was Agile, providing my customers with more focused processes and products.

Having said that, you might have the same question as many others who have asked me whether the methodology alone will help to deliver the best product without rigid quality processes in place. Agile has in itself quality as a core principle. This methodology can derive the best output.

I have found that process power can be more powerful that even position power. You may know managers who have significant position power but who cannot get their staff to work effectively towards a common derived goal. Process power saves the day every time. However, all this does not mean that a laissez-faire approach is the modus operandi when it comes to leading core teams. There are times when consensus is best and times when a directive style is the way to go. Process power is your way out of the storming and the chaos that comes out of it. However, having said 'process is power,' the software development processes we are institutionalizing today are more about bonding artifacts to a certification rather than improving the product quality. When valuable output is not foreseen, project teams simply consider the process as good to go for compliance rather than practice in the project for bettering the productivity. Finally, the majority of customers also feel the same. They consider the processes to be an overkill of time, gaining no special benefit. In an onsite/offshore-centric model where the majority of the teams reside at the offshore location, process is followed closely but is not transparent to the customers.

However, there was another side of the coin that worked well for me. Well-applied good process works miracles. Seeing yourself as managing a process that in turn leads the people will be a powerful new paradigm that I have experienced. Blending the Agile methodologies with quality processes has a very successful track record in which the customer has attributed the process adherence to the engagements following the Agile methodologies. This in turn has provided gains to the engagement not only from the Agile perspective but also from the quality processes. This was also well received by my internal and external auditors (ISO, CMM, etc.) who had evidenced these projects as 100 percent process compliant.

Further in this chapter, I will discuss some of the implementations in which I was successful in coalescing my organization's quality processes with the Agile methodology. Though it was my first engagement marrying both Agile and quality processes and convincing the quality leads of the value of the tailored processes, once everyone agreed, this approach ws used for all the rest of my assignments. I have tailored the processes and documents for each of the life cycle phases, evidencing with the most widely accepted processes by the teams and the organization.

Capturing User Stories

Requirements capture the intended behavior of the system. As we know in Agile terminology, requirements are denoted as user stories and user stories are written down many times in a hard copy format and sometimes in a soft copy format.

Fig 8.1: *Story cards scanned and stored at a central location on the server*

Whichever way you capture the user stories, ensure that they are scanned (if the user stories are in a hard copy format) and stored away at a central location that will form the base to be defined as a "requirement document/repository" for the project. Since Agile methodologies enforce an iterative approach, storing the user stories when finalized has proven very useful to me in terms of managing the changes as well as process compliance. In the majority of my engagements, all the user stories, after finalization, were scanned and placed in the server as the finalized versions of the user stories. Figure 8.1 depicts the server repository for the user stories in one of my assignments on Agile. This repository served as proof to the SQA teams on the coverage and documentation of the requirements for the assignment.

Design

Agile does not try to avoid creating design documents. In fact, design is mandatory for any project and not a complementary work product. Whether you intend to create design or not, the moment you write a line of code, you are realizing a design. It may be in your head, but it is still a design. If you are on a team of professionals, then to really "scale" communication, a written and agreed-upon design is a very helpful (I would say "necessary") task on the path to success.

Personally, I have also found that by writing out a design, I have found errors and holes in my own thinking. I have uncovered serious and major risks just below the "architectural" level, deep in a component design. Sometimes this risk will make or break a project!

Whatever life cycle model you choose, the underlying design concepts are the same. What I have been practicing is to avoid creating the design strategies for each project but to use the already defined principles and design guidelines followed for other projects within the organization, tailoring it to my project requirements. This has avoided creating comprehensive design strategies and guidelines during the design phase.

Whether a high-level design or low-level design, it should always be tagged to a user story or a logical group of user stories. Whichever way you have captured the design, logically it should be grouped together in some form or other that will become a final set of design documents for the project.

What we had been doing on one of the Agile–based projects is to design for each user story created in the Rational tool from which we generated the design documents at periodic intervals.

In another engagement, the design was drawn physically on cards for each user story and the coordinator took a scan of these cards at regular intervals to paste in a consolidated document indicated as the "design document." However, this methodology is best suited only for small engagements that are simple in nature where complexity does not play a major role in the design approach.

Both ways are effective and can be selected based on the size of the project along with the tools used in your organization.

Code Development

"Single code base" should be our slogan, not only for Agile development but for all development projects. It is mandated that for any project with an iterative approach, a configuration management tool should be institutionalized that will serve the purpose of avoiding code overrides and mismatches. Multiple code streams are an enormous source of waste in development of a code base. I fix a defect in the current version of the code base. Then I retrofix the same change to all other deployed versions and the active development branch. Then you find out that my fix has broken something else in your code and you interrupt me to fix my fix. The story goes on and on.

There are legitimate reasons to have multiple versions of the same code base active at one time but the plan should be to reduce the multiple versions gradually. Whatever you have to do, improve processes so that you no longer need multiple versions of the code. The configuration management tool itself would lead us towards process compliance.

Testing

Creating comprehensive test strategies for a release will lead to extended work force time utilization. Hence, you should institutionalize usage of the already created test strategies from similar kinds of engagements and review them to create the strategies for your project.

Creating exhaustive test cases for unit testing, system testing, etc. would also be a tedious exercise for an Agile–based engagement. What should be imposed for the testing teams (and development teams for unit testing) is not to create extensive test cases but to test the system first in the right way. A tester should completely understand the user story before thoroughly testing the user story. What has worked best for me in one of the engagements is that the testing teams brainstorm with the developer on the user story and write down the sequence of the test steps for each of the user stories. The steps were captured on the paper card and were finally scanned to store in the repository for further evidence.

In another engagement over the period of maturity, we implemented the practice of Automated Unit Testing by using some of the industry-accepted tools. It helped the teams to split Test First versus Automated Tests. Start with Automated Tests, and then add Test First when the teams are comfortable in the first iteration. Once you have Automated Tests in place and a framework with which you are comfortable, it is much easier to get into test-first situation.

Whatever tools or methodology you implement, a test leader should ensure and perform period checks to ensure the completeness of the testing for each of the user stories. Figure 8.2 depicts a tracking mechanism that has been used by my test leaders to track the progress of testing for each user story or, more precisely, the sub function for each user story. This has also proved the SQA team's adherence to their processes on the test coverage.

155

User Story #	User Story Name	Sub Function Name	Test Executed	Result	Re-testing Needed	Tester
US 001	Create Training Request	Add New Training Request	Yes	Fail	Yes	Gopala
US 002		Modify Training Request	Yes	Pass	No	K T R B Sharma
US 003		Delete Training Request	Yes	Pass	No	G Murthy
US 004	Create Course	Add Course	Yes	Fail	Yes	Vijay Raghavan

Fig 8.2: Test tracking

Change Control Strategy

"*Changes to requirements are always welcome*" is the slogan that follows in an Agile development approach but the key expression behind this slogan needs to be understood clearly. Changes to requirements for any user story should be accepted anytime before the start of the design phase or, at the most, before the start of development for specific iteration of the user story. If we have crossed any of these boundaries, the change in a user story should be considered as a change request that will follow the same procedure of a new user story starting its phase from estimation until deployment.

Make it a practice to track all the change requests in a central location so that you will be able to assess the volatility of the requirement changes. You can use a simple tracking sheet for all the change requests. Figure 8.3 depicts the tracking procedure that was used by my teams in one of the Agile–based engagements. This has proven a very useful tool in tracking change requests.

User Story #	User Story Name	Sub Function Name	Base Configuration	Base Configuration Date	New Configuration	New Configuration Date	Change Log
US 001	Create Training Request	Add New Training Request	US 001 - Create TR ver 1.0	12-Mar-07	US 001 - Create TR ver 2.0	22-Mar-07	1
US 012	Create Course	Add Course	US 012 - Create CS ver 2.0	12-Jan-07	US 012 - Create CS ver 3.0	12-Feb-07	2
US 100	Approval Workflow	Approve Training Request	US 100 - Approve TR ver 1.0	1-Jan-06	US 100 - Approve TR ver 2.0	12-Oct-06	1

Fig 8.3: Change Control Tracking

Estimation and Effort Tracking

You can measure your development process only by tracking the estimated vs. the actual effort. In project development, it is common to experience differences of 50 percent to 100 percent deviations in the estimates for a project, even when the bidders are using proven project estimation techniques. Some people believe that software development will, by its nature, defy our efforts to measure it and if that is what you believe, then it is my job to convince you that you are wrong, especially with the Agile model.

In an Agile process model, you should derive the estimates for each story card using a work breakdown model or any other proven methodology that best suits your assignment, where you can estimate the effort required for each phase of the life cycle. If the story card tends to be a bigger piece of functionality, try to break it into smaller parts (called subfunctions) and then accumulate them to arrive at the final estimate for a user story. This will be your best approach to an accurate estimate. However, your experience in estimating for the initial set of user stories will also form a base to arrive at a better estimate for the next set of user stories. In fact, people get cynical when estimating projects, often taking a best guess and multiplying it by two or three, which definitely should not be your approach, according to my experiences.

This has been the traditional approach to IT projects that have deviations in their estimates, at least minimally. But to draw the correct conclusion, you need to track the actual efforts spent. With the Agile model of software development, what I have experienced is that when the requirements are broken down into smaller sets of user stories, the chances of huge deviations in terms of effort spent is much less, and with proven techniques and experiences, the risk of large deviations is drastically reduced.

With the current trend in the industry of many distinctive and complex tools being used by every organization to track the efforts of each individual resource, it has slowly become a baffling and a prolonged exercise for the teams to record their actual efforts with these tools. This, in turn, has created discomfort within the teams who would rather use these tools for compliance than accuracy. My recommendation for people who read this book, is to reduce the boredom entailed in capturing actual work effort by each team member into the high complex tools (with lots of additional data to be keyed in apart from the efforts) which will result in productivity losses on the valuable project work to be done. This does not imply that you should stop tracking the efforts, however.

Sub Function Name	Estimated Effort (PH)				Actual Effort (PH)				Deviation					Effort Deviation
	Analysis	Design	Coding	Testing	Analysis	Design	Coding	Testing	Analysis	Design	Coding	Testing	Overall	
Add New Training Request	4	8	12	6	2	16	12	6	-20%	+50%	0%	0%	+30%	
Add Course	2	16	6	3	4	8	3	3	+20%	-50%	-50%	0%	-80%	On Track
Approve Training Request	6	6	6	6	6	6	6	6	0%	0%	0%	0%	0%	

Fig. 8.4: Effort Tracking

    Have a simple tracking sheet (refer to Figure 8.4 which has been used to track the teams' efforts in one of my Agile–based engagements) which will be updated by each team member at the end of the day that takes not more than five minutes for each resource. Alternatively, the leader can approach each team member individually to collect the data in a hard copy format and then can input the details into a tracking sheet.

    This type of tracking sheet has proven very useful to me during the execution of a project where the lessons learned on the deviations have been used when preparing the estimates for the next set of user stories. In addition, if demonstrated wisely, the tracking mechanism serves as proof for estimates and effort tracking. What I have always tracked for my engagements is the man-hours expended vs.

estimated which demonstrated the time it took to complete each user story (not just billable time), thereby rolling out the complete effort expended for the entire project including the deviations. It religiously reflects the total hours required to create the application, including testing, rework, removing defects, administration, and documentation effort. I carry these experiences forward as lessons learned for the next set of engagements to better the productivity and the estimates of the next engagements.

Budget Tracking

Cost tracking plays a very vital role for project and program managers; in fact, for all mid-level and senior managers in an organization. How can you tell if your project is on budget? Use Microsoft Project™ to compare original cost estimates, actual costs, projected costs, and see the variances between costs at any time and at any level of detail. I will not get into the details of what the cost tracking merits and demerits of the product are but I will provide insight on simple ways to track the budget of your project, thereby focusing more on the project deliveries.

The best option to track your budget is to use planning tools such as Microsoft Project. Derive the total estimated cost by entering all the necessary data such as pay rates, per-use cost, fixed cost, costs for tasks, resources, etc.

After the project begins, you update task progress— the amount of work done on tasks or the percentage of the tasks that are complete, from which you can derive the costs for your project based on task progress.

By combining the actual costs of completed work with the estimated costs for remaining work, you can derive the scheduled (projected) costs. More importantly, the difference between the scheduled and baseline costs can be derived. It is this difference, or cost variance, that tells you whether your project is on budget or not.

You can do simple cost tracking by viewing the actual and scheduled (projected) costs for tasks, resources, assignments, and the project as a whole.

To determine whether you are on budget or not, you can view the cost variances between scheduled costs and baseline costs.

Defect Tracking

Apart from the strategy of continuous development, what I foresee in an Agile development methodology is continuous improvement.

Defect tracking is a fundamental and critical part of application lifecycle management. Within a project, one of the most important tasks to manage effectively in order to ensure delivery of functionality by the end of iteration is the management of defects, which involves the identification of defects, the assignment of defects to the appropriate owners, and resolution in a timely and efficient manner. Outstanding defects can hold up the release of functionality to the business, denying them the benefits of the project and damaging your reputation for being able to deliver an effective product on time.

The fundamentals of effective defect tracking are the same, whether you are using pencil and paper, an Excel spreadsheet, or a full-fledged defect-tracking tool. Implementation of some of the following fundamentals provides a solid foundation for success:

- **Information Capture**: Guidelines should be established that determine the minimum amount of information necessary to report a defect. For example, determine what information is needed on descriptions and screenshots so that a developer will be able to reproduce the defect and fix it. Defect tracking must be simple enough so that teams will use it effectively but should not be oversimplified because it must capture vital information about the defect.

- **Communication**: An open line of back-and-forth communication can facilitate constructive dialogue between who reported the defect and who is responsible for fixing it.

- **Versatile Environment**: Functions should be tested with the production environments. To provide thorough testing, the team must identify and test all possible hardware/software combinations.

To support this, what you should be implementing is a rigorous defect management process with the bare minimum functionality to raise and track the defects to closure.

Figure 8.5 depicts the spreadsheet that I used in one of the projects, placed at a central location accessible to all the developers, testers, and management teams in a shared directory.

User Story #	User Story Name	Sub Function Name	Owner (Developer)	Coding	Tester Name	Bug Description	Bug Type	Assigned TO	Status
US 001	Create Training Request	Add New Training Request	Gopal Krishna Behara	Completed	K T R B Sharma	When I click the submit button after entering all the details, it displays a blank page	Major	Gopala	Pending FIX
US 012	Create Course	Add Course	Gopal Krishna Behara	In Progress	Not Assigned				
US 100	Approval Workflow	Approve Training Request	Gopal Krishna Behara	Completed	G S M R Murthy	Training Request name is not displayed with ""	Minor	Gopala	FIXED & Assigned for Retesting

Fig. 8.5: Defect tracking

In this tracking sheet, the project manager maintains the data and assigns the responsibilities for testing each user story to the testers. Testers will be able to access and update the testing status of each story assigned and appropriately inform the development teams via a defined reporting process on the status of testing. The critical factor of this approach is to ensure very close interaction between the development and testing teams so that there is no time lag in communication created by the respective managers.

These defects are finally consolidated and an analysis is performed to break out the reason for these defects. This analysis is then carried out as lessons learned in my next engagements.

Release Management:

With scores of people involved in iterative development using Agile methods, it is of utmost importance to have a clear release plan in place for each iteration and share it with the teams so that the transparency on the dates for each phase and each iteration are clearly understood by the executing teams and the customer. A release plan classically incorporates aspects such as duration of each iteration along with the duration of each phase within the iteration apart from the number of resources required for iteration or a phase. Figure 8.6 depicts the release plan my teams used in one of my assignments.

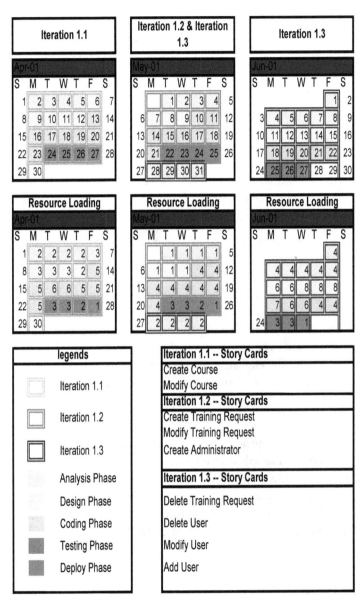

Fig. 8.6: Release plan

This spreadsheet forms the evidence for the schedule given to the SQA teams as well. As a first step in this activity, all the user stories should be prioritized, based on mutual agreement between the customer and the project teams. After this, the story cards should be grouped together to define an iteration, or more specifically, a release. The outcome of this activity at the end should be a complete release plan consisting of all the requirements pertaining to the project and the timelines to their completion. However, based on the project needs and further clarity on the story cards, reshuffling of the user stories between iterations is permitted and the schedules for iterations can be altered from time to time.

Let us turn now to my other innovative experiences, which I derived after my first implementation of an Agile project in the late 90s. As mentioned earlier, "best practices" and learning carried forward to the next engagements provides cases for a better execution model in any engagement. The learning I had in my first implementation was an exhaustive list of documents and spreadsheets that my project teams had to maintain for the engagement. Though maintaining these was considered 50 percent more time saving than the disciplined use of QA checklists, for me, 50 percent was also a large percentile.

Then, I turned to implementing an Agile tracking tool that has a simple user interface and all of the checklists mentioned above can be tracked within the tool itself. This provides an easy way for the teams to maintain all the data in a central repository. I named the tool "My Agile Tracking Tool." Refer to Figure 8.6 for a snapshot of the tool.

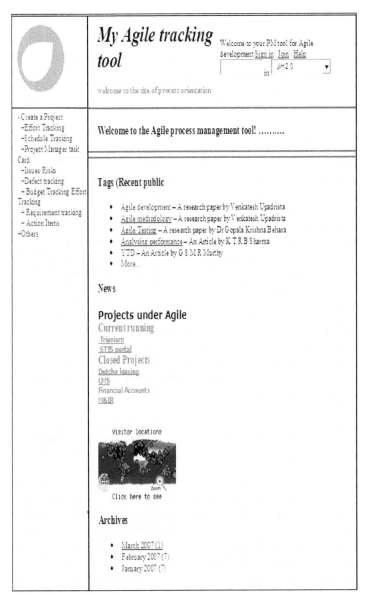

Fig. 8.7: My Agile project management tool

This tool has united the project management activities with the tracking of requirements, design documents, tests, and defects so that everyone directly or indirectly representing a project has a real-time picture of the project's features and quality, priorities, roadblocks, and risks. "My Agile tracking tool" has deep support for all roles and activities in the development lifecycle and has led me to replace some of the expensive integrated point tools with a hosted Web interface that starts quickly and scales easily across platforms and geographies with low risk and low cost. Even my customer had access to some of the sections to track the progress of the project.

My next project implementations executed Agile using this tool and had a high satisfaction ratio by the project teams (users). A user satisfaction survey was conducted on the tool after the end of the project execution and I had an overwhelming response. Some of the responses to the survey were:

"In this tool, I only enter data which is useful for me."

Gopala Krishna Behara – Junior Developer

"Tool is fantastic. I need not spend huge amounts of time entering the data, as this is similar to entering and tracking my project in an Excel sheet."

G S M R Murthy – Senior Developer

"This tool gives me lots of reports! Fantastic tool."

Chakradhar S – Project Manager

"I am overwhelmed with the work done by you and your teams. These tool and processes are very beneficial to achieve a productive Agile business value. This tool has convinced me that you and your teams are with me on my journey towards understanding how Agile can help me develop applications and services. You guys rock!"

Customer Manager

I am not recommending that everyone should go out and create an exhaustive tool to track an Agile–based engagement. What I have learned from my experience is that having all the tracking templates in a central database will help any organization gather metrics from projects, which can be helpful for any new assignments initiated under a similar background. In addition, there are numerous tools available in the market for Agile process management. You can explore some of these to see if there is a match for your organization.

However, if you and your organization are comfortable enough gathering all the necessary metrics and tracking the project using your own sophisticated tracking spreadsheets, my advice to you would be to continue using them. A tool should not be an encumbrance to management and the project teams.

CHAPTER NINE

Case Studies

I t is easy to get caught in the rigor of executing your projects under the Agile model losing sight of the original motivating and feasibility factors behind instituting those practices in the first place.

To illustrate the theory of Agile in practice, I have collected a few of my experiences on how I have approached the Agile software development methodology. This case demonstrates how I was successful during the initial days of project execution using Agile for a leading bank (one of our privileged customers.) I used Agile to help solve several customer dissatisfaction issues:

- Longer release cycles for a couple of large scale projects in the account

- Our inability to deliver a function in a shorter time period

- Our inability to absorb change

It all started in the late 90s when my initial engagement (on Agile) started with this customer—a major bank in Japan. My organization had been servicing this customer for almost five years. I entered this engagement as the relationship manager and I had the privilege of meeting the top management teams (along with the senior and mid-management) from the customer organization. During mutual discussions with some of the business and technology leaders, I sensed a bit of discomfort in terms of our organization's execution model, but overall, they seemed to be happy with the kind of services provided. That *"bit of dissatisfaction"* worried me enough to find the root cause, which I felt otherwise might have a ripple effect leading to bigger issues and larger audiences in the future.

I slowly began my investigation into the details of the execution model and my steps began with interviewing each of the business and technology teams to understand the different perspectives on their inputs to the currently executed model. My questions during the interview mainly focused on two key sets of properties of the engagement dynamics.

My first set of questions was aimed at understanding the customer satisfaction ratio on the basic parameters of delivery. Some of the questions were as mentioned:

Q1) What is your satisfaction level with the quality of the deliveries that are being serviced by my organization? If you were to give feedback on how we have improved from the year we started our relationship with you, how would you rate us on a scale of 1–3 (1 being not satisfied, 2 being satisfied, and 3 being extraordinary performance)?

In answer to this question, the majority of the customers said: today, if you ask me this question,

I am satisfied with the quality of your deliveries. However, we had some issues that cropped up during the initial days of the engagement. Over the intervening period, you have fixed them all; however, we are looking for better output by greatly reducing the number of defects so that my business is happier with our ongoing relationship.

Q2) Did you have any instance where my organization had major effort deviations from your expectations? If you were to give feedback on how we have improved from the year we started our relationship with you, how would you rate us on a scale of 1–3?

In answer to this question, the majority of the customers said, deviation was a concern in the earlier days, but no issues for me now.

Q3) Did you have any instance where my organization had major schedule variations with regards to your expectations? If you were to give feedback on how we have improved from the year we started our relationship with you, how would you rate us on a scale of 1–3?

In answer to this question, the majority of the customers said, deviation was a concern in the earlier days, but no issues for me now.

Q4) What is your satisfaction level with my organization's capabilities to absorb changes during execution of the project? If you were to give feedback on how we have improved from the

year we started our relationship with you, how would you rate us on a scale of 1–3?

In answer to this question, the majority of the customers said: "I would not say this is a concern, but you have to improve on this."

The outcome of these sessions was an average rating of 2 for the first set of questions.

My next questions were aimed at understanding the resource competencies onsite and offshore as well as on the management teams.

Q1) What is your satisfaction level with the quality of the onsite resources provided to you in the engagement?

In answer to this question, the majority of the customers rated us with exceptional performance, apart from the few who said satisfactory.

Q2) What is your satisfaction level with the quality of the offshore resources provided to you in the engagement?

In answer to this question, the majority of the customers said, good resources but need to improve communication. They have a high level of commitment.

Q3) What is your satisfaction level with the quality of the management teams onsite?

In answer to this question, the majority of the customers were highly satisfied.

Q4) What is your satisfaction level with the quality of the management teams offsite?

In answer to this question, the majority of the customers noted that we had good teams. Very few rated us as unsatisfactory.

The outcome of these sessions was an average rating between 2 and 3 for the second set of questions.

This was not the end to my research. I immediately called off my time at the customer location for a week and traveled to my offshore center. There I spent extensive time gathering all the offshore teams' feedback, informing them of the feedback from the customer, and getting their views. I pulled out all the customer satisfaction surveys for each project in the engagement to find out their ratings for each year on each engagement.

Finally, after thorough analysis, interviewing sessions with my onsite/offshore managers and project teams, I finally drew my conclusion on the findings. The findings are depicted in Figure 9.1.

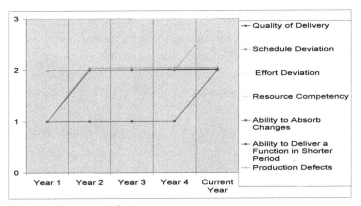

Fig. 9.1: Customer satisfaction survey

The conclusion of my research was that the majority of my customer teams were extremely happy with the competency of the resources deployed for the engagement but had a low satisfaction level in terms of the quality of the deliverables and the delivery model.

The quality of the deliverables was at the expected value but we had never done an extraordinary job of exceeding customer expectations. This resulted from the number of defects raised by the customer during UAT for each release. Though defects for each release were within the acceptable limits, we had not drastically reduced them over the period. What I traced, hashing out things with my project teams, was that there were several instances where the customer had changed requirements during the later stages of development, which was not within the time limits where we could accommodate them. In addition, some of the teams said that the customer frequently changed the requirements.

Do you see a ripple effect here from requirement change management to defects and finally leading to customer dissatisfaction?

Dissatisfaction of the customer with respect to my organization's *inability to deliver functions in shorter intervals* was another area of concern that came out of this research. My project teams demonstrated how some of the customers were demanding to test the software at regular intervals to get a comfortable feel for the functionality of the release, which they were not able to satisfy due to the life cycle model they followed.

Are you not thinking now—that I said I entered the engagement in the capacity of a relationship management role but performed a delivery role? You have to understand one thing—that no organization can have business opportunities in this world of competition without having set the customer satisfaction parameter very high. Therefore, what I believe is to set the engagement in the right direction first and gain the utmost confidence of the customer to grow the business into a mutually collaborative model.

Now, my research had come to a standstill with many data points and my initial findings based on the parameters of the customer's dissatisfaction. My next step was to have a process in place to address the long-lasting issues. In due course, I recognized the need for Agile implementation for some of the projects in the engagement to overcome the systemic problems. I enlisted help from my delivery teams to identify the right set of processes to target and the right set of people who would have a favorable influence on team spirit by placing group objectives above self-interest. Finally, I decided upon an Agile process model for this engagement, which I felt was the best match.

With all the Agile principles focused on resources, process and tools, iterative development, customer availability, etc., I sorted out the most relevant principles that should be used in the delivery model of this engagement. I am not negating the fact that for an Agile–based development model, we have to account for all these principles in order to have a successful implementation, but the gist of solving this current issue in the engagement revolved only around the definition of the delivery model. The other principle regarding resourcing did not play a major role as we had the right set of motivated and strong resources within the projects. We then defined an iterative model for the project delivery in which we break the development into smaller sets of iterations and all SDLC phases follow for each such iteration.

However, there were complications with this model during our engagement. Here are a few of them:

1. The production systems were in the hands of another vendor and having an incremental deployment for each project in a duration of one

to two weeks would be too costly to the customer as the other vendor woudl charge extra fees to support this approach.

2. The business users' availability around the clock for the project teams was another issue.

3. The processes and tools used in my organization cannot be tailored to my specifications overnight.

4. The teams working on the projects were not accustomed to the Agile development model.

I targeted the easiest of the issues first, which was making the team aware of Agile principles, concentrating mainly on what they gain by using this model, which would give me buy-in from my teams. I always believe in acceptance rather than enforcement. Only if the team feels that a process can make their lives easier and will have good credibility within the organization, will they provide their complete cooperation, and Agile is known for this. Therefore, I started with a basic "crash course" of Agile training and made the team aware of what changes they might see in the near future. The session focused on the painful areas for the customer as well as the execution teams, highlighting the deficiencies that created a need for the Agile process implementation. This was very well supported by the project teams as they saw how the process could make a difference in their mode of execution.

The next challenging task for me was to tailor my processes to match the agility of the project using these methods. This was easy for me as we had some examples of Agile project implementation in the organization and samples of how tracking sheets were used as an alternative to the process-related tools satisfying all the audits. I had a discussion with the QA teams, tools group, etc. that turned out to be quite receptive to my effort to adopt the Agile development model.

Now, the teams did not have to spend much time on the tools and processes but only had to input the necessary data. Saying that to my teams made them much happier as they saw improvements right at the start of the engagement using Agile.

The most challenging task was to play an effective role with the customer making them understand the advantages of the new process. Policies that originate from the internal employees within the customer organization carry the advantage that they will certainly be heard by top management and accepted or at least modified, but certainly not rejected if they are for a good cause. However, in a vendor relationship, this is much more difficult, as the initial impression which I give my customer for this change will be the judging factor for the innovation. My approach was simple. I took my proposal to the most prominent and powerful technology director within the customer organization and explained all the merits of the new process. After a prolonged discussion, we finally agreed to pilot test the proposal on one of the projects. This definitely was the opportunity for which I had been looking, and I was confident that my model would succeed. I was now able to convince my business users of the new model and that their availability was required to the teams for the Agile model. However, the important proposition was also that business users provide proxies so that my team could always be in touch with the proxies to get information in a timely manner. The use of proxies was also important, so that business users need not spend 100 percent of their time on the engagement, but rather budget their time to perform a UAT for each iteration. The users accepted this very well. In the end, they wanted a good product without wasting much of their energy.

Now, the next hurdle—how to plan the deployment of iterative releases without any additional cost. This kind of problem always arises for a service provider. Lastly, we decide to use our integration environment for having a pre-UAT performed for each iteration. This approach carries a risk factor, as the data in an integration environment is not exactly the same as in production; however, we were left with no choice because to build trust with the customer, we have to prove ourselves and win. Therefore, the plan was to have a pre-UAT for each release and, after the end of the final iteration, have a random check on all the functions considering a completed UAT phase.

Eventually, we had all the process and approvals in place to kick start the new release. I had been incessantly monitoring the progress of the release and we had the best to implement it. The continuous progress of the release led to its becoming, to the best of my belief, the most comprehensive program of its kind in the account, leading to a satisfied customer. (Perhaps another reason was that Agile was becoming popular during that crucial period when the ideological foundations of the Waterfall model were falling apart.) The release had a six–month duration of which each iteration was planned for two to three weeks. After a prolonged battle, my team was able to complete the first iteration build for pre-UAT and the results were very positive with no defects. One reason for this was the small functionality delivered, duly respecting and highly acknowledging the most important aspect: my team members' hard work and their drive to do their best.

We now set about to look at the next set of iterations, replan, and produce the desired outcome. Of course, we started with the expansion of my Agile rollout and with 'selling the idea.' There was one important aspect I discovered during my time in this experience—the qualities I found in my leaders. Having said so, what are the characteristics of the person who best leads in a complex problem-solving environment where he or she is supported by an able set of executives? This seems a straightforward enough question. However, a straightforward answer is not viable without one clarifying the question further. Is the best leader the one who is most acceptable to the group and the group sets the leader as an example for themselves, with the personality, behavior, and image that is most fit for what people seek in a leader? Or, is the best leader the one who is most likely, during the tenure of his or her work, to enable the project to reach its goals via the teams? That important difference exists between leaders, and is well attested to in research literature. Unfortunately, elected leaders are not necessarily effective in achieving their goals. If I am left with the choice of selecting between these two sets of leaders, my choice from a management standpoint is only clear: the effective leader has to be chosen who best motivates his or her team by setting an example.

Moving back to my discussion, we had finally passed the six months of the project cycle where we delivered almost twelve iterations. We were finally successful in deploying the release into production, and the team was awaiting the customer's response.

Now to head off the gathering storm clouds, I moved faster to get the results. So far, so good. My initial impressions, based on the environment at my customer location the next morning, were assuring me of success. I finally moved into the conference where all the stakeholders of the project were present, each having a copy of the so-called "customer satisfaction survey" template with him. The customers greeted me in the room with smiling and welcoming faces, reassuring me and confirming that it had been a successful deployment. All the stakeholders had given high satisfaction ratings to the project and we moved to a very successful release. Figure 9.2 demonstrates the results. It was a big success; however, all the credit goes to my execution teams.

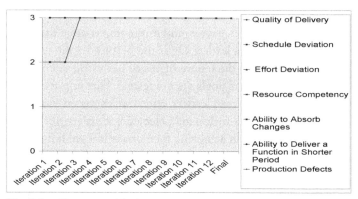

Fig. 9.2: Customer satisfaction survey

This was definitely not the end to my story. I carried forward the experience to the top customer executives and stakeholders to demonstrate our capabilities and the success story in this engagement. I wanted to regulate the model for all the projects with this customer. After a long review process, I was able to convince all my stakeholders to form an Agile PMO within the organization and mandate that all projects implement the Agile processes that meet the defined criteria. In the end, we had the PMO in place and we were seen as successful leaders in Agile methodologies.

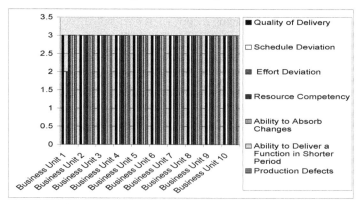

Fig. 9.3: Customer satisfaction survey for the whole engagement

After some time, we had to travel on short notice to Minneapolis for another business opportunity, but since the processes were in place, I had no need to assist further with this assignment. When I returned a couple of weeks later, I saw that the engagement was humming along and we had a high satisfaction rating from the customer. In other words, we had achieved utmost customer satisfaction. After a year, we collected the engagement satisfaction survey which indicated a great success. Figure 9.3 depicts the survey results for all the projects. They are grouped by business unit for ease of representation, as we had many projects executing within each customer business unit.

You have to note the point that not all the projects were transformed to the Agile development model. Careful analysis of customer expectations and the life cycle models being used led to certain projects implementing the Agile methodology and the rest of the projects following the Waterfall development model.

The final lesson out of this experience was to prepare for success with the right goal and thorough analysis, overcoming hurdles as they occur. Do not protect yourself from success by holding back. Do your best and then deal with the circumstances. That is Agile. You leave yourself exposed. For many people, this is extremely scary, but for some, it is daily life. That is why there are such polarized reactions to Agile.

Now, how do teams such as the ones described above, which have emerged as consistently effective over the years in our experimental studies, compare with the teams in the most successful companies in the business world? Do the same underlying principles and patterns apply, or do they not?

Answers to these questions are not easy to provide. Certainly, successful companies can be identified by their record of performance. The main difficulty lies in finding an adequate means of describing how these companies are run and in describing what processes drive them.

Nevertheless, some useful insights can be gained by looking closely at a very few companies which have features relevant to the theme of my enquiry.

CHAPTER TEN

Leader's Orientation
Towards Optimization

Optimization is all about utilizing the available resources efficiently and effectively to the best possible extent. These resources can be humans, processes, hardware, software, or a combination of all.

You might have the best breed of the resource and the best of the processes in your organization but overall, if you are not able utilize them to the utmost, your organization will not reap the benefits. Some organizations have made huge investments in the best resources but have not yielded the desired returns on these investments.

Let's take a simple example in which you have just driven a new Mercedes out of the showroom which has all the sophisticated software and hardware to give you a high performance output. It has everything installed—innovative tumble flaps to give you a high fuel efficiency, a 6–cylinder diesel engine, and the GPS software which will guide you with directions to go anywhere in the country.

In this case, the driver is the leader and the different parts of the car are the resources. My driver always knows the basics of driving more efficiently.

- He avoids aggressive driving (speeding, rapid acceleration, and braking) which lowers the gas consumption.

- He avoids excessive idling.

- He uses cruise control.

- He uses overdrive gears, etc.

In this way, I use my car very efficiently to obtain the best possible output. In addition, the utilization of my car is high, as I drive almost 200 miles per day as compared to others who use the same car at the rate of 10 miles per day. In this manner, I assume that I get high utilization of the car that I purchased.

My driver also uses the GPS facility to drive to different locations, which is the process I have set to avoid any wrong routes. In this way, I ensure that I reach my destination effectively,

In this way, my car gives the best mileage and my leader utilizes the resources of the car to derive the best output and use it efficiently.

Though I had said that my utilization is high by using the car effectively, how was I able to arrive at this conclusion? I had a process set up to measure these factors in which

- I determined the average number of miles per month to find out whether my car utilization was good.

- I looked back to see whether I had taken any wrong routes during the duration to gauge my effectiveness.

- Finally, I calculated the average mileage to ensure that I had used my car efficiently.

Utilization is the term defined to gauge how much you are using out of what you have. Efficiency defines the way in which all the skills are utilized to the best possible extent to derive effective output.

Hence, optimization defines the mixture of utilization and effectiveness to derive an effective output.

Optimization is all about these two principles of utilization and effectiveness as demonstrated in Figure 10.1.

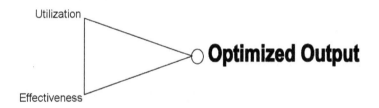

Fig. 10.1: Utilization and effectiveness applied together derives an optimized output

As an example, consider a classic example of failure in which my manager Chakradhar had deployed Gopal Krishna on a project demanding X technology. Gopal was highly skilled in Y technology but also had only a fledgling knowledge of X technology. Because of the high demands of the customer, Chakradhar had no other choice than to deploy Gopal for the opportunity. Another reason was that Gopal was a favorite of Chakradhar's. In this experience, though Gopal was being utilized completely, he was not able to effectively provide the desired output due to his weaker skills in the technology, resulting in an optimization that was not acceptable.

Low effectiveness thereby translates to more wastage of effort to obtain the desired output, which proved true in Gopal case. Though Gopal had demonstrated his hard working skills to drive the project to success, the extra hours over the weekends he had put forth could have been avoided by deploying the right technology resource match.

If you are able to achieve the defined project goal on time and with quality, why does anyone need to bother about the efficiency?

The answer is quite simple. If you can achieve a desired output with x effort and you foresee no reasons to spend x+ effort, then you not are ambitious enough for the account growth that translates to organization growth. Many organizations have failed to digest these basic principles and have not proven successful in the end. Do you want to be one of the successful leaders?

In managed services Agile projects—the onsite/ offshore model—this tends to be the most important factor in terms of resource efficiency and utilization. If these factors are compromised at any point in time, you will fail during the initial iterations themselves. The reason is that you will tend to spend more time working on a specific iteration, which might be the case in which you do not run out of schedule but target either resource frustration from working overtime or having a higher resource load than what the situation actually demanded. Both are quite risky for the long-term sustenance of the engagement as deploying more than the required number of resources will overshoot the budget, and making the resources work overtime will hurt morale.

To maintain success with the customer and the organization, optimization to the best possible extent is in order, and optimization is all about two parameters: utilization and efficiency.

So, who is responsible for getting these parameters right? Is it the team leader, the manager, the senior manager, top management, or the CEO of the organization? The answer to this is that there is responsibility at each reporting level within the organization and the control should flow from top to bottom—only then can you consider the account to be optimized. Figure 10.2 demonstrates a simple case of account optimization in which each line in the reporting structure has the responsibility to optimize their resources going down the chain.

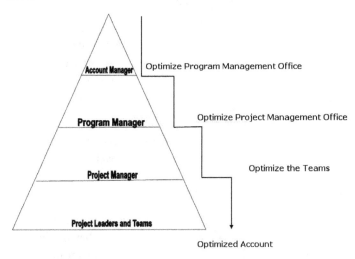

Fig. 10.2: Optimization at the account level

In this case, the account manager holds the overall responsibility to optimize his account and the responsibility flows down the line to each manager and subordinates for their respective projects.

An account manager here tries to apply the principles of utilization and efficiency to optimize his program management office by defining the right set of program managers to yield an effective output. The goals are set for the program management office to optimize the project management office and so the flow goes.

Typically, the same principle also applies to the organization as a whole, where each individual is responsible for their area of work and performs the optimization at their own level, translating the responsibilities to the next level of hierarchy. Figure 10.3 demonstrates the flow.

As demonstrated above, if the rules of utilization and efficiency are applied in the right spirit, you can achieve the best possible results of optimization.

Having looked at the principles of optimization, how do we really measure whether we have already optimized the account/organization or if it needs further optimization? This is not a simple task, as for each level, metrics need to be gathered and careful analysis has to be performed to understand the overall level of optimization.

Utilization is all about assigning resources to the work and efficiency is the measure of how well the assigned resources are producing results and whether the right skilled individuals are assigned to the right work.

One measure of utilization is to ensure that all the resources are assigned to work. However, does having the resources assigned to their respective work really mean that the account is optimized? *Definitely not.*

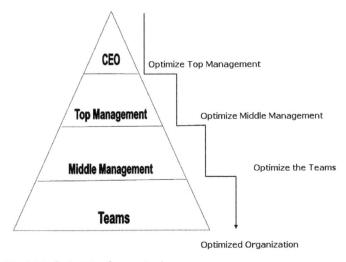

Fig. 10.3: Optimizing the organization

In many organizations, utilization is directly measured by the billing ratio. If a resource is 100 percent billed in a specific project, then the resource is considered as utilized. Though it might be considered true from the organization standpoint, the fact is that billing does not translate to optimization. This high utilization figure affects the buyer and the seller's inability to perform more. If the customer is smart enough, then this kind of engagement will not last for long since the same services are offered by a competing vendor at a lower cost (primarily because they do not consider utilization as a direct translator to billing).

Let's get into detail to understand the concept more.

What does productivity actually mean in relation to software development? The simple answer is that productivity is an expression of the amount of work that is done (or is planned to be done) during a certain period. The more work a resource does per unit of time, the better his or her performance. Put differently, the productivity of a resource is measured by the number of input (data) units it takes to transform into output (data) units in a given time. Moreover, at any given point in time, a resource deployed for a set activity performs the best if he has the primary skills in the performing area of work.

As an example, a resource with X skills can give the best output if he is deployed in the assignment demanding X skills. If, due to various influencing factors, he is deployed in an area not matching his X skills, his productivity will be low, thereby impacting the overall optimization. This is similar to a software algorithm requiring 20 steps to be taken before one knows where new data is to be inserted. This has greater impact on a program is performance than a database setup that does the same in 10 steps.

One of my experiences was where we were executing a multi-million-dollar engagement for a major bank in the United States that was very understanding and had huge resource growth. The engagement initially started as a staff argumentation engagement and slowly transformed into a managed services contract requiring much of my team's capabilities to execute the engagement. During the course of the engagement, resource demands were serviced immediately and that was one of the contributors to our success.

I was operating as the account lead for the engagement and the engagement was humming along with a lot of revenue in the pipeline forecast. What had been the differentiator for me from other competitors was the case of Agile in the projects, and my team's faster resolution time to provide resources and make successful deliveries. My managers always reported utilization of 80% and above to me for their respective projects but these figures were based on the billing of the resource. It then occurred to me that billing should not be the only factor that is considered for the best utilization. During those days, I had been researching a lot on optimization concepts and the profound finding that hit me was, "Why are we considering billing as the mode to calculate the utilization, when it is definitely not the right way to determine the utilization?" With this principle in mind, I thought there should be something more that needs to be done.

I then started to apply my concepts of optimization to the engagement.

I initiated the collection of a lot of metrics on the resources deployed into the engagement, considering billing least important. After a long and thorough exercise, I was able to draw up the findings as shown in Figure 10.4.

What data I collected for this exercise is mentioned below:

- Number of resources at each role deployed in the engagement

- The required skills at each level for each project

- The skill sets of the resources deployed in each role

- The percentage of resources that have the skills that match the "project required skills"

- The percentage of resources that have the skills not matching the "project required skills"

Roles	Number Of Resources	Project Required Skills	Resource Available Skills	% Of Matched Skills	% Of Un-matched Skills
Account Head	10	Account Management	Account Management	100%	0%
Program Managers	20	Program Management in Java Technologies	Program Management in Java and MS Technologies	50%	50%
Project Managers	32	Project Management in Java Technologies	Project Management in Java and MS Technologies	30%	70%
Technical Leads	60	Java,J2EE, SOA Framework, Application server	Mix of MS technologies and Java Technologies	90%	10%
Developers	160	Java,J2EE, SOA Framework, Application server	Mix of MS technologies and Java Technologies	70%	30%
Jr Developers	270	Java,J2EE, SOA Framework, Application server	Mix of MS technologies and Java Technologies	60%	40%

Fig. 10.4: Current executed engagement statistics

After collecting these data points, I set about to collect the metrics of productivity for resources based on the competency level in the desired technology. I set about to draw a conclusion on the figures of productivity for resources that had the primary skills of the project. Figure 10.5 and Figure 10.6 show the results.

	Number Of Lines Of Code/Day					
% Of Matched Skills	**100%**	**80%**	**60%**	**40%**	**20%**	**10%**
Technical Leads	508	436	240	160	84	64
Developers	457.2	392	216	144	75.6	57.6
Jr Developers	406.4	349	192	128	67.2	51.2

Fig. 10.5: Productivity factor of development teams

	Activities/Day					
% Of Matched Skills	**100%**	**80%**	**60%**	**40%**	**20%**	**10%**
Account Head	100%	100%	100%	100%	100%	100%
Program Managers	100%	100%	100%	100%	100%	100%
Project Managers	100%	80%	60%	40%	20%	10%

Fig. 10.6: Productivity factor of management teams

From Figure 10.5, a technical lead with 100 percent of matched skills can deliver 508 lines of code per day while a resource with 40 percent of matched skills can deliver 160 lines of code per day.

Similarly, a junior developer with 100 percent of matched skills can deliver 406.4 lines of code per day while a resource with 40 percent of matched skills can deliver only 128 lines of code per day.

This clearly determines that a resource to be deployed in the engagement should have the right set of skills required for the project to deliver the right kind of output. So if you can achieve better productivity having the right matches of resources, why do you need to deploy a low proficiency resource on the project? The answer might be two-fold. One is where you are not able to find the right resource and technology match in time to meet project demand. The other is where you assume that since the resource deployed thought the required skill is not a primary skill for him, he has little knowledge on the specific skill to work on the project. Both are not acceptable. However, we will discuss this in detail later in this chapter.

The next step in my analysis was for a different set of findings on optimization, which was at the resource composition for each role level. This was quite uneven to achieve optimal optimization in the current operating engagement.

If you refer to Figure 10.7 carefully, you will find out that there is a lot of management overhead on the account. Figure 10.7 details the percentage of allocation at each role level.

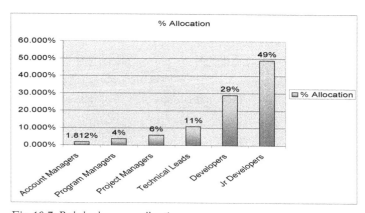

Fig. 10.7: Role-level resource allocation

If you observe Figure 10.7, you can find out that out of a 596 person team, on an average

- A team lead is managing an average of 3 resources

- A project manager is managing a team of 15 resources

- A program manager is leading an average team of 29 resources

After collecting all these data points, I set about to plan for optimizing the account. Though I was sure that this was a massive exercise and having the changes done in a single shot would lead my engagement to failure, I set about to tackle the dynamics slowly—step by step.

There were two principles that I had to set right in the engagement, which had to go hand in hand. One was to have the resource allocation right in terms of the numbers and, at the same time, ensure that the right skilled resources for the project were available and deployed to achieve optimal productivity.

The first action for me was the take a bottom-up approach in which I slowly started to understand the work performed by my direct reports—my program managers. The task of ramping down the program managers for me was simple enough since from Figure 10.6, you can clearly understand that I already know the technology-resource match. A program manager in my organization holds a very senior role and has the capability to handle a team of 60 resources, but the way my current assignment was executing, these managers were managing numbers less than their capacity. Therefore, wherever I got a chance to deploy a program manager for my other accounts, I started moving the managers from this account into the other, promising some greater roles and responsibilities in the new arena. Slowly, I started loading my existing program managers with more roles and responsibilities, providing them higher opportunities to grow in responsibility. I started about setting the same goals for them to have an optimal level of their direct reports in the account, i.e., the project managers, and so, the responsibilities and goals translated to the lower levels in the hierarchy. If you refer to Figure 10.6 once again, my program managers had to set their minds to the important principles of optimization, which was to first roll off the staff who were not the right technology match for the project and next to roll off more staff to achieve the level of optimization.

The most challenging task for me as well as my managers was the changes to the resource allocation at the bottom level of the account, i.e., at the level of technical leads, developers, and junior developers. Since the number of people involved was high, changes had to be done slowly. How I planned to tackle this situation was a step-by-step process that I will now desdcribe.

For projects that are moving to closure and that demand a change in resources due to skill-set mismatch:

1. Identify the resources from this project who were deployed in the projects having a different primary skill set than that required by the engagement, i.e., many of the projects in the engagement demanded J2EE skills, but there were a couple of resources having the core Java skills that lacked expertise in J2EE technology. Also, a couple of resources had the primary skills of the Microsoft technologies and had little knowledge in the Java/ J2EE technology, but due to the project demand, these resources were deployed on the project anyway.

2. At the end of the project, provide training to the existing resources that lack the J2EE skills but who are highly proficient in the core Java skills

and whose leaders believe that these resources can scale up very easily, provided they undergo proper training.

3. Replace resources not having the required skills for the project and redeploy them in other accounts/projects that demand Microsoft skills. One important point to be noted here is that I do not condone the practice here of direct replacement of team members who lack a specific skill. Careful analysis has to be performed on the resource strength. If you feel comfortable with the idea that a specific resource, though not strong in the currently required skills now, has the passion to learn the skills, then this would be the right resource for your project. Give him or her the time and the support they need to learn the required skills, and you will observe that in the end, they will be active contributors to your project.

4. Identify replacements with the right skills.

5. Deploy the new resources for the assignment.

Moving back, the changes mentioned above were closely coordinated with optimizing the count of the low-level staff as well. The optimal level of team dynamics that has worked best for me is mentioned below:

- A team lead should be able to manage a team of at least 6–8 developers.

- A project manager should be able to manage a team of 12–15 resources on average.

- A program manager should be able to manage a team size of 60–80 resources on average.

Since the count definitely will differ, based on the complexity of the application and other influencing factors of customer expectation, overhead, etc., these numbers should not be taken as benchmarks for your engagement. Analysis has to be performed at each level for your engagement to arrive at the optimal number of resources that each level can handle; however, this was the right mix for my engagement.

After a long stint of almost six to eight months, I was able to improve the engagement. Though we had faced many challenges during the course of these changes, which is outside the scope of this discussion, we were on our way to a successful engagement.

Roles	Number Of Resources	Project Required Skills	Resource Available Skills	% Of Matched Skills	% Of Un-matched Skills
Account Head	1	Account Management	Account Management	100%	0%
Program Managers	10	Program Management in Java Technologies	Program Management in Java and MS Technologies	100%	0%
Project Managers	33	Project Management in Java Technologies	Project Management in Java and MS Technologies	80%	20%
Technical Leads	60	Java, J2EE, SOA Framework, Application server	Mix of MS technologies and Java Technologies	98%	2%
Developers	130	Java, J2EE, SOA Framework, Application server	Mix of MS technologies and Java Technologies	90%	10%
Jr Developers	320	Java, J2EE, SOA Framework, Application server	Mix of MS technologies and Java Technologies	87%	13%

Fig. 10.8: Engagement statistics after changes

Refer to Figure 10.8, and you will be able to draw the conclusion that I was able reduce the resource count of the total engagement by 8 percent which translated to 8 percent productivity improvement. In addition, the engagement now had the right skilled resources that in turn engaged the engagement to have good productivity across the board.

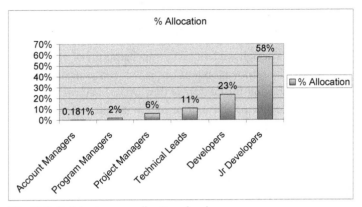

Fig. 10.9: Role level resource allocation after changes

Refer to Figure 10.9, and you will be able to determine that utilization at each level of the hierarchy was distributed evenly, having each individual perform his responsibilities to his utmost.

The benefits derived from these changes are mentioned below.

- Due to having resources with the right technology skills match, each individual was more productive in his or her area of work. This in turn improved team motivation as the work was distributed evenly and nobody had to work overtime to keep his or her tasks on schedule.

- Higher (and desired) responsibilities were given to each leader giving him or her ample scope to perform more and demonstrate his or her capabilities, as well as grow in the account and in the organization.

- Utilization was not now a direct representation of billing and customer satisfaction improved as they saw more work performed with fewer resources.

This gave us the chance to penetrate more into the customer organization and grow our business.

- Though we had lost a small amount of revenue by removing resources, it was compensated by having the customer flow other business to us, believing that our productivity was a competitive differentiator.

Slowly, in the due course of the engagement, we were able to achieve 90 percent optimization in the account. The same principles are now considered as best practices in my organization, slowly absorbing into the organization level benchmarks.

Though I have described this new approach in simple terms, this was not an easy task to achieve, as it involved many pushbacks from the stakeholders and the project teams, and I had to include many compromises and initiate escalations to get the engagement to the level demonstrated above. Therefore, the only way to avoid this kind of stress for you as the leader is to plan the engagement right during the initial days and have the right set of principles during the start of the engagement. If you have understood the concepts I have demonstrated above, you, as the leader, will not have to run around later to set the engagement right.

The right process is to identify the right kind of resources up front and not have to reduce them later due to project pressures. A customer can bear the loss of delay but would not like to bear the loss of productivity. If you are able to understand this simple fact, you will be a winner for a long time.

Many leaders increase the number of resources with the increase in number of inputs to be worked on. In plain terms, the more work pours onto the plate, the greater the resource count goes. Though this concept is acceptable, fair judgment should be used in adding resources, confirming the fact that the team is optimized before adding more resources. The number of resources is not directly proportional to the output generated, and there are cases in which more resources are added to a specific inputwith no change (or an overall reduction) in output.

Let me share a simple experience highlighting this issue. We had been working on a time-critical application, executing an assignment for a very prestigious and demanding customer. We had very strict SLAs governing the schedule and deliveries for the assignment and the agreement was that any changes or additions to critical functions would be accepted and delivered within the customer-defined timeliness with very few negotiations. Since our relationship with this customer was new, our chances of negating the SLAs were minimal, else we would have lost the relationship. Though the timeliness was clearly governed by the SLAs, I had the confidence within myself that my negotiation skills would play a major role in resolving any dire situations. Therefore, we started the engagement with a positive attitude and had the best of my leaders deployed in the assignment. We had maintained a pool of resources that could easily take care of any additional needs of the project.

We were able to delivery the first iteration successfully with high customer satisfaction; however, during the course of the engagement, issues started cropping up in terms of customer dissatisfaction with a deviation from the SLAs. I started to understand the iteration dynamics and the mode in which the changes or additions to the requirements were taking place.

Whenever there was any new functionality added to an iteration, resources from the free pool were deployed immediately and were made to work on the deliveries. Though this worked to some extent, when the count reached a certain threshold, the output came to a standstill or decreased. Figure 10.10 shows the results. For simplicity, I refer to each individual function of the requirements (or story cards) as inputs and outputs.

Input (Number of Requirements)	Duration of the Release (In Days)	Number of Resources	Achieved Output (Number of Functions)
20	9	8	20
30	9	8	30
40	10	9	40
50	10	11	50
60	12	13	60
70	12	15	70
80	12	17	73
90	12	19	73
100	12	21	70

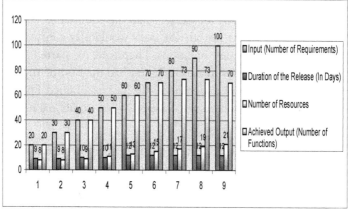

Fig. 10.10: Number of inputs vs. the output

If you observe Figure 10.10 carefully, there was an expected amount of output generated until we reached a level of 70 input points, and gradually, though there was an increase in the number of resources, there was no considerable improvement in the output. The reasons are three-fold as mentioned below.

- Optimization principles had not been applied before ramping up the resource count. When inputs increase, first try using existing resources to find out whether the team is really optimized and all the members are equally loaded and have the right skills to achieve a productive output. If there are some deviations, first try to optimize the teams and then plan for resource ramp-up. I have experienced that an optimized, self-sustained team generates good output even in cases where the inputs have increased with no resource ramp-up. However, this is not a generalized principle and has to be applied on a case-by-case basis.

- The leader failed to understand the concept that an increase in the number of resources is not directly proportional to the output generated. Reasons being:

  ▸ Work cannot be distributed evenly among the resources if the inputs cannot be logically separated.

  ▸ Interdependent inputs cannot be distributed (i.e., cases in which the output of one input forms the input of another function).

  ▸ New resources brought on board cannot be 100 percent effective from day one.

  ▸ Existing resources have to spend some amount of time training new resources.

- The leader failed to understand the concept that to generate an optimal output, an increase in the number of inputs should be proportional to the increase in the number of resources, also accounting for the proportional increment in duration.

So what was the contributing factor to the failure in the iteration demonstrated above? Was it due to the team's capabilities or was it due to bad planning? The answer obviously is bad planning.

My next step was to ensure that my leaders understood the dilemma in proper planning, especially for engagements that demand an iterative approach, and it provided them the necessary support to plan the next iteration. I also had a discussion with the customer to explain to them the reasons for the deviation in the initial iteration. Here I negotiated for flexibility on the schedule for iterations that had the likely chances of requirement changes or additions, and the negotiations worked well for me with the customer.

We started to commence the next iteration after investing in planning time with my leaders. As a first step, my leaders had to understand the complexity of each request (here called inputs or, in Agile terms, the functions of a story card) and determine the effort required for each such input. Figure 10.11 demonstrates the sample effort breakdown that each resource would expend for one input at each level based on his or her role.

Input Type	Person hrs of Effort (Technical Leader)	Person hrs of Effort (Developer)	Person hrs of Effort (Jr Developer)
Simple	8	9.6	11.52
Medium	16	19.2	23.04
Complex	24	28.8	34.56

Fig. 10.11: Effort for the inputs based on the complexity (sample)

The next step for me, based on the number of inputs for the iteration, was to determine the optimal number of resources required at each level. Based on the resource load and the estimates, we were able to determine that this iteration would require a minimum of eight resources to complete the 37 inputs determined for this iteration. (A discussion on how I arrived at the resource load is outside of the scope of this discussion, assuming that you are aware of how to derive resource load based on the estimates.) Out of the eight resources:

- One was deployed in the capacity of technical leader

- Two were the developers

- Five were the junior developers

Based on the total number of inputs, the number of resources, and the schedule for the iteration, we were able to prepare a work distribution for each resource. Figure 10.12 demonstrates the work distribution between the different resources. One interesting point that has been captured is that the utilization figures at each level (which is of utmost importance) attain an effective optimization for the team.

The figure clearly demonstrates that the average utilization in the team is around 80 percent and there is scope for absorbing at least another 15 percent. Using this tracking mechanism, my leaders were able to agree that my approach to utilization and the effective mode of resource loading was effective.

Input Type	Technical Lead		Developer		Jr Developer	
	Number of Inputs	Efforts Required	Number of Inputs	Efforts Required	Number of Inputs	Efforts Required
Simple	0	0	2	19.2	16	184.32
Medium	2	32	4	76.8	9	207.36
Complex	2	48	2	57.6	0	0
Effort Required for 1 Resource	10		19.2		48.96	
# Of Resources in the Team	1		2		5	
Person Days(PD) of Effort for all Resoures	10		9.6		9.792	
Total # of Days for the Iteration	12					
Utilization @ each Level	83%		80%		82%	

Fig. 10.12: Work allocation and utilization figure at each role level

However, utilization had not been limited to specific roles and my leaders had gone beyond my initial strategy to prepare utilization figures for each individual on the team. This had been very helpful to them since they were able to determine which resources were optimally loaded and which were not, allowing them to be able to load the unutilized resources in case of emergency.

Figure 10.13 demonstrates the resource utilization figures, which clearly depict that a couple of resources on the team were not completely loaded.

Apart from this, the dependencies between the different inputs had also been determined while performing the work distribution so that effective loading could be performed based on the interdependencies.

Figure 10.14 depicts the interdependent components and Figure 10.15 depicts the work distribution of the inputs.

Role	Resource Name	# of Inputs Assigned	Complexity	Utilization Period (Person Days)	Utilization
Technical Leader	K T R B Sharma	2	Medium	12	83%
		2	Complex		
Developer	Gopala Krishna Behara	1	Simple	12	70%
		1	Medium		
		1	Complex		
	G Murthy	1	Simple	12	90%
		3	Medium		
		1	Complex		
Jr Developer	Paul	6	Simple	12	100%
		2	Medium		
	Erik	4	Simple	12	82%
		2	Medium		
	George	4	Simple	12	82%
		2	Medium		
	Kishore	2	Simple	12	60%
		3	Medium		

Fig. 10.13: Resource wise utilization figures

The advantage of this approach is that a resource need not have to wait for the code of other resources to be completed to start his own code, and vice versa. All dependent functions are assigned to a single resource (or group of resources) who work closely with each other and, in these cases, lost time is minimal.

Next came another case where the customer asked for additional inputs to be developed within the duration and the planning for these inputs was two-fold: first, 11 inputs were added to the iteration and a detailed feasibility analysis was performed to determine

1.   Whether the new inputs were interdependent with the existing inputs for the iteration

Total Number of Inputs	Tightly Coupled Inputs (Groups)	Loosely Coupled Inputs
	i1, i3	i11
	i2, i5, i4	i12
	i6, I7, i8, i9, i10	i13
	i16, i17, i18, i19, i20, i21	i14
	i25, i26, i27	i15
	i28, i29	i22
37	i35, i36, i37	i23
		i24
		i30
		i31
		i32
		i33
		i34

Fig. 10.14: Dependencies between inputs

Resource	Inputs Assigned Considering Dependencies	Utilization
K T R B Sharma	i1, i3 i28, i29	83%
Gopala Krishna Behara	i11 i12 i13	70%
G Murthy	i14 i25, i26, i27 i15	90%
Paul	i16, i17, i18, i19, i20, i21 i22, i23	100%
Erik	i2, i5, i4, i24 i33, i34	82%
George	i6, I7, i8, i9 i10, i30	82%
Kishore	i31, i32 i35, i36, i37	60%

Fig. 10.15: Work distribution based on the interdependent functions

208

2. Whether the resources available would be able to take the load of completing the additional inputs during the iteration

Let's assume that the additional inputs are numbered from i38 to i48. Here is the analysis of the dependencies of these inputs.

- i38 is dependent on i1, and i3

- i39 is dependent on i12

- i40 is dependent on i25, i26, and i27

- i41 is dependent on i15

- i42, i43, i44, and i45 are dependent on i31, and i32

- i46 is dependent on i10, and i30

- i47, and i48 are dependent on i33, and i34

With these dependencies in mind, these new inputs were distributed among the existing teams as depicted in Figure 10.16. Items marked in **bold** are additions to the existing resource load.

However, we were able to produce the best output with the existing team keeping utilization in mind. If we had not derived the utilization factor from the start of the iteration, we really would not have been in a position to absorb the additional work with the existing teams. This would have led to a situation where we add new resources to the existing team, thereby causing the team to spend time training the new resources and requiring additional monitoring from the leader, which would have led to additional efforts from the existing teams.

Resource	Inputs Assigned Considering Dependencies	Utilization
K T R B Sharma	i1, i3, **i38**	92%
	i28, i29	
Gopala Krishna Behara	i11	92%
	i12, **i39**	
	i13	
G Murthy	i14	98%
	i25, i26, i27, **i40**	
	i15, **i41**	
Paul	i16, i17, i18, i19, i20, i21	100%
	i22, i23	
Erik	i2, i5, i4, i24	90%
	i33, i34, **i47, i48**	
George	i6, i7, i8, i9	90%
	i10, i30, **i46**	
Kishore	i31, i32, **i42, i43, i44, i45**	94%
	i35, i36, i37	

Figure 10.16: Work distribution based on the interdependent functions

In another case, we had to absorb additional inputs during an iteration where the interdependencies between iterations were not permitting the work to be absorbed easily. The utilization of the resources reached a peak, and it was determined that any additional requirements could not be absorbed by the existing team. In this case, we asked to extend the schedule and increase resources. The reason was because the new inputs were highly dependent on the existing ones, and distributing these to the new resources would affect interdependencies. Therefore, we determined that resources working on the prior inputs should work on these new interdependent inputs. For this to work, there had to be a definitive change to the schedule (Figure 10.17). At the same time, resources were increased to help meet the delivery schedule. Items marked in **bold** are the changes made to the schedule, inputs, and the resources.

Old schedule & resource loading			
Resource	**Inputs Assigned Considering Dependencies**	**Duration (Person Days)**	**Utilization**
K T R B Sharma	k1, k3, k38	7	92%
	k28, k29	3	
Gopala Krishna Behara	k11	3	92%
	k12, k39	2	
	k13	5	
G Murthy	k14	1	98%
	k25, k26, k27, k40	6	
	k15, k41	3	
Paul	k16, k17, k18, k19, k20, k21	7	100%
	k22, k23	3	
Erik	k2, k5, k4, k24	6	90%
	k33, k34, k47, k48	4	
George	k6, k7, k8, k9	5	90%
	k10, k30, k46	5	
Kishore	k31, k32, k42, k43, k44, k45	6	94%
	k35, k36, k37	4	
New Schedule & Resource Loading after Requirement Addition			
Resource	**Inputs Assigned Considering Dependencies**	**Duration (Person Days)**	**Utilization**
K T R B Sharma	k1, k3, k38, **k49, k50**	10	92%
	k28, k29	3	
Gopala Krishna Behara	k11	3	92%
	k12, k39	2	
	k13	5	
G Murthy	k14	1	98%
	k25, k26, k27, k40	6	
	k15, k41	3	
Paul	k16, k17, k18, k19, k20, k21	7	100%
	k22, k23	3	
Erik	k2, k5, k4, k24	6	90%
	k33, k34, k47, k48, **k51**	**6**	
George	k6, k7, k8, k9	5	90%
	k10, k30, k46	5	
Kishore	k31, k32, k42, k43, k44, k45	6	94%
	k35, k36, k37	4	
New Resource 1	**K52, K53**	**3**	80%
	K54, k55, k56	**7**	
New Resource 2	**k57**	**2**	60%
	k58, k59	**4**	

Fig. 10.17: Comparison between the original and new schedule and resource load

Finally, after thorough planning, and considering the impact of utilization, we were able to put the engagement back on track and were highly appreciated not only by the customer but also by my top-level stakeholders in the organization for implementing the concept of effective optimization. This has given me a strong mandate to apply the same concept in other accounts and deploy it throughout our organization.

What has to be understood from the experience I have mentioned above is that utilization is crucial to any success, and a careful analysis of the engagement execution will point the way to success. The only factor to be considered for better productivity and the factor to be emphasized is that the right kind of utilization leads to optimal output. If you refer to Figure 10.18, you can clearly see that productivity increases with utilization, and if you try to increase utilization beyond a certain point, the productivity decreases, as overworked people are less efficient.

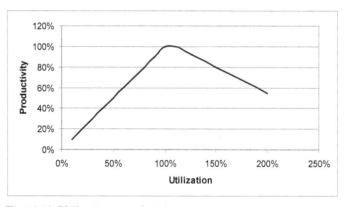

Fig. 10.18: Utilization vs. productivity

Keep these points in mind and you will be successful. As a leader, your job is not just to monitor the work performed by the teams and act as a liaison between the customer and the teams—you have more on your plate. A successful leader is one who understands that the responsibility of a leader is to understand and work on the principles of effective utilization apart from the general team and project management activities. If these basic principles are understood and implemented, they clear the way for a smooth execution of the project, while leaving the team with high morale since none of the members of the teams will be overloaded or underutilized.

If you have thoroughly understood the concepts of optimization discussed above and the need for leaders focused on optimization, you may have considered the following points:

1. Having resources in the team that do not possess project-specific skills as their primary skills underutilizes them and, at the same time, does not provide a chance to utilize their skills for other projects that need these skills. In a way, you are not only impacting yourself but also others in the organization.

2. With a forecast showing more work in the near future, many leaders over invest or build buffers in the project; this also reduces utilization again, having an impact on yourself and others in the organization. Also, since it is you who has invested, you will definitely not be interested in releasing your newly-trained resources to other opportunities.

The only solution to these issues is careful planning, which will not only help you as an individual leader but also will lead the rest of the staff to better results.

You should try to strike the right balance between supply and demand. I am not saying that you should avoid creating buffers but the target should be to utilize these buffer resources within a short span, rather than holding on to them long for future assignments.

What I have learned from this study is that one should be able to utilize effectively the available resources to their utmost capabilities, which will drive the best output for your engagement. Ramp-ups are definitely tolerable but only when the utilization within the existing teams has reached its peak. Refer to Figure 10.19, which shows how one of my teams responded to me, focusing on this optimization concept.

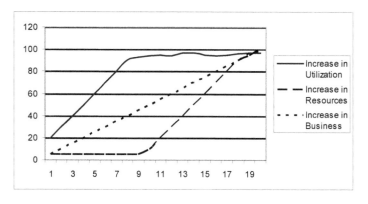

Fig. 10.19: Optimization concept for effective utilization

From the figure, you can clearly see that with an increase in business (or in plain terms, an increase in work) ramping up the teams was not planned. The first strategy applied was to consume effectively the available resources until utilization reached its peak, and after the threshold, the ramp-ups were started.

What this shows is that an increase in revenue or projects should not be directly proportional to the increase in resources, and these factors should not determine your ramp-up plans. Effective utilization of existing teams should be the first game plan to absorb the work, and only then should the the ramp-up plans be implemented.

On a similar note, I recall another experience in which I was able to influence the HR policies of recruitment for the future business. Though it had not directly translated to utilization improvements, the procedures I had adopted helped me to have the right resource ramp-ups at the right time, thereby maximizing the utilization. These policies applied to recruitment of the senior level staff, but the same principles could not be applied to the low-level staff due to the different complexities involved.

Here is the basic principle: after my business development managers have responded to a request for proposal and have clear indications that we are likely to win the engagement, my recruitment efforts begin. I determine the probable start date of the engagement, and based on that as a safe target, I start my recruitment well in advance. I try to find the right match inside the organization, using many internal references, and if finally I am not able to find the right match, I go for external recruitment. For external recruitment, I follow some other basic principles. The step-by-step approach is

1. I inform my recruitment teams to sort out the best possible résumés from the market with those having worked in the required technologies and domains as the preferred ones. I provide them a decent lead time so that I can have the best resources available for interviews. My steps of initiating the recruitment process only start once the bid is submitted and I have clear indications that the bid will be won. From the proposed bid, my teams determine the composition of the teams at various levels and determine the number of resources needed.

2. I clearly instruct my resourcing teams to communicate to the interviewees that the review process will require quite some time due to an account policy describing the defined rounds of interviews. It was not hard for my recruitment teams to convince the interviewee, however, as my organization had great credibility in the market. Parallel to step 2, my bid managers continuously follow-up on the status of the bids submitted.

3. After a couple of rounds of interviews, suitable candidates are short listed. We now start the game plan. I discuss the status of the bid with my bid managers, and if we have a likely chance of initiating the engagement, we start the paperwork for the selected resource since we will have lead time for the resource to join the organization during which the project will slowly pick up momentum.

    ▸ If for any reason, the bid results are likely to be delayed, we try to negotiate a matching delay in the resource paper processing and keep the candidates on hold.

    ▸ If we have lost the bid, we either push the candidate for other customers for their requirement or terminate the recruitment but we always keep these resources in the pool so that we can recruit the candidate for future possible positions, based on the resource's availability.

Figure 10.20 depicts a detailed step-by-step process of the recruitment process.

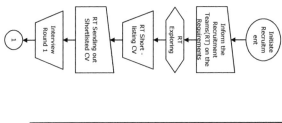

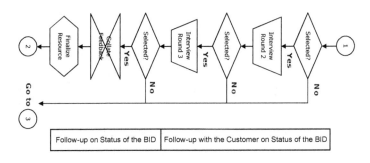

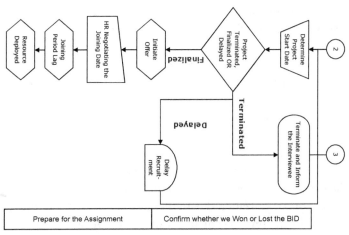

Fig. 10.20: Flowchart for my external recruitment

If you refer to the strategy in Figure 10.20, you might feel that this is not a fair judgment on the candidates, as I initiate recruitment with unconfirmed data. However, this process has definite advantages not only to my organization and me but also to the candidates as well.

In many organizations, recruitment happens based on the unconfirmed proposals and there are several cases in which the resources are recruited for a position, which really does not exist once the candidate joins. From an organizational perspective, it leaves the resource unutilized until any matched positions are sorted out for this resource, and during the course of this identification, there is a major loss in utilization. Though there are lots of norms governing the utilization for newly-recruited resources during the initial days, you cannot use them as a guideline to determine if a resource is under utilized.

With the model I have implemented. I make hiring decisions based on an accepted proposal, which will result in much less under utilization once the resource is hired.

On the other hand, many senior resources join new organizations with the zeal to overcome the weak spot they encountered in their previous organization and perform better in the new organization. If these resources are not provided with the right kind of assignment at the start of their career in the new organization, they tend to lose confidence with the organization, moving them into an uncomfortable zone. I have experienced many cases in which resources joining an organization have been demotivated when promised roles and work were not assigned to them during the initial days with the new organization. There were many cases of employees leaving the organization due to these experiences. Therefore, I only go by the principle of better utilization of new resources from their first days in the organization. Why do I want to have the wrong recruitment strategies in place that not only inconvenience the new resources but also the organization as a whole?

Though candidates might feel discouraged for some time not getting into the organization, this definitely will not be greater than the regret they will feel after joining a new organization that has no defined work assignments for them.

Having discussed so much about the right skill deployment on projects and utilizing these skills for generating any optimal output, will these factors alone lead to an effective optimization? These are definitely many of the influencing factors for optimal output but there are some other dimensions that are related to people practices.

In general, however good your planning is, there are two definite parameters a leader has to ensure when deploying resources to generate the best performance: the potential of the resource, and the resource's readiness to work. There are different influencing factors that can decrease the 'passion to work,' or readiness of resources, that should be taken care of by the leader. Some of them are mentioned below:

- Work interruptions

- Other resources misleading the flow of work

- Increased wait times

- Low motivation of the resource

- Lack of challenges within the assignment.

You have to understand the simple fact that every individual wants to succeed but, in a majority of the cases, the right kind of opportunity never hits their door.

"Passion to work" is defined as the ability to work with high enthusiasm, motivation, and in the right direction, and passion is very dynamic and changes based on the environment in which the resource is working.

Though many leaders understand and implement these optimization principles, expecting a high performance from their resources, a hit on the 'passion to work' parameter will reduce performance, sometimes to a level that you will be unable to gauge. Figure 10.21 represents the different combinations of passion and potential and how they affect output. In this figure, 1 denotes low, 2 denotes average and 3 denotes high.

If you take an example of one scenario from the figure in which the potential is high but the ardor or the passion is low, the resource generates an average output. A desired high-performance output is generated only when the potential and passion both are high or in cases where the performance is average but the passion is high.

However, if either of the two parameters of potential or passion is completely zero, the output generated will also round to zero. Hence, you should never have a team either zero on the potential or zero on passion.

As a leader, you can ensure that the 'passion to work' within the team is high by providing the right kind of environment:

1. Ensure that the teams are happy with the work they are performing.

2. Ensure that the teams have an uninterrupted work environment.

3. Ensure that other team members are not blocking the flow of work or giving wrong directions.

On one project, one of my developers was working on code that had to undergo a design deviation with X and Y

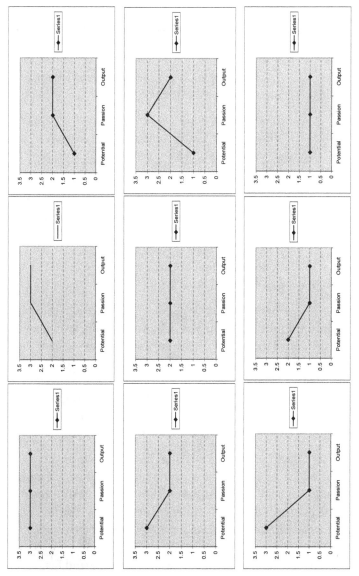

Fig. 10.21: Output generated with combinations of passion and potential

alternatives. While this resource was pretty sure that X was the right direction, based on another person's input, the Y direction was chosen. It turned out that X was the right choice, and there was major rework at the end of the release.

1. Ensure that people get the perks, recognition, and rewards that they deserve.

2. Ensure that people are not overloaded with work. This indirectly affects their passion to work.

3. Ensure that people need not wait for inputs to accomplish their work. In other words, give the team a uniform flow of work so that they do not get broken up, etc.

These are just a few of the factors that can lead to high motivation and passion to work.

Leaders always try to get the best-of-breed technical resources on the team with the right skills for the assignment, but the dominating factor that many miss during the course of the engagement is the measurement of the passion of the team. A resource may be highly passionate at the start of an engagement, but during the course of the project, due to different influencing factors, his or her zeal to work may go down, lowering performance.

Having discussed much on how the performance of an individual is highly dependent on his or her potential and passion, how do we now obtain that best performance and, in the end, how does each leader ensure that the individual's performance reaches expectations? If you refer to Figure 10.22 which gives you a consolidated view of output based on potential and passion, you can clearly understand that in cases where the potential is average and passion is high, this also leads to high performance results. The reason is that if an individual has the zeal to perform more, with hard work an average performer can move into the high-performance zone, but a high performer with 'low passion to work' will always produce a poor performance.

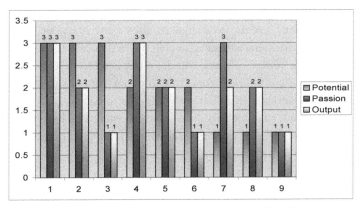

Fig. 10.22: A consolidated view of output based on potential and passion

So, encourage resources to move from low or average performance levels to high performance by developing the passion to work and to win. Provide people with the right environment in which they can perform to their best and you will see a visible difference, even in cases of resources with low potential. You can definitely change people's passion more easily than their potential.

One more important principle that you have to digest is that you will only have 20 percent of your team being top performers. Do not assume that you will have the best and the top performers on your team from the initiation of the assignment; this assumption is completely wrong. The principle you have to opt for is to *build* high performers on your team rather than *hiring* them into your team. As a leader, you have to motivate the team and provide them the directions to better their performance.

It is true that you have to identify top performers, but many leaders concentrate on the top so much that the mighty middle and the low performers are often ignored. If these groups are neglected, then you will lose almost 80 percent of your team, which will never lead to future success.

Many leaders, even organizations, concentrate on the top performers, actively ignoring the middle and the low performers. With A players, you definitely need to motivate them more, but the B players are also important contributors to success and you should be providing them equal opportunities to grow and show their potential. You have to go the extra mile in figuring out the motivating factors and be equally fair to everyone.

The consequences of such benign neglect can include excessive turnover, high attrition, reputation damage, and project disasters. You need to know who your top performers are, and at the same time watch the message you are sending, because disenfranchised middle and low performers can be just as damaging to your business as the loss of all of your top talent.

You have to understand the weak areas of your low and middle performers and give them the support and direction to generate improved performance and move them into the high performance zone. Though this is not easy, it is not impossible.

Identifying the Right Kind of Supervisors

Who is a successful supervisor? Here are some of the attributes found in the best supervisors.

The Desire to Do More

A successful supervisor is one who pushes the work to be done under his very close supervision but at the same time pulls back the work in cases that demand it, thereby balancing the pressures on his team. He or she plays an effective role utilizing resources to gain maximum productivity. He or she does not view the status from the cockpit of an airliner at the 30,000-foot level but gets into the details to understand the day-to-day operations. He or she is not content with delegating the tasks and responsibilities to the next level in the hierarchy; rather, he or she bypasses the next level and gets into the details. Results are an area of motivation and he or she continuously strives to achieve results, which is a differentiator respected highly by his teams and his supervisors. He or she has a high level of self-confidence within himself and always strives to achieve more than that defined for him. He is the one who accepts the facts and works hard towards the goals to achieve. Good supervisors have desire to achieve great heights with their hard work, dedication, and accurate planning. I call these leaders taskmasters.

The Ability to Lead

Wise and successful supervisors understand the fact that they cannot be successful without their teams being successful. They always try to gain the best productivity from teams through optimization. They provide their teams with ample opportunities to demonstrate their skills by putting them in the highly-optimized zone to deliver higher productivity. They apply the principle of team motivation to themselves and act as successful leaders. They help their subordinates to feel part of the team and enhance morale, cooperation, productivity, and quality. They make their job easier by having the decisions made together with the teams so that there are no conflicts later. They are the ones who review ways to conduct more productive (and shorter) meetings and who will discover the wonder of synergy. They are the ones who get along with their teams to understand the team complexities and try to solve them collectively, providing the team a sense of ownership and togetherness. They are the ones who develop the STARS in the team. It is often said that a good leader develops his or her own replacement. Successful leaders are not afraid to develop those beneath them, as emerging leaders will attest to how good of a leader he or she really is.

Successful leaders are those who provide their teams with the right kind of support and direction to ensure success. They do not hesitate to provide negative feedback to the resources themselves, providing them ample opportunity to improve, and guiding them to overcome their personal challenges. They provide the right kind of feedback at the right time, not waiting for the feedback to be shared at the end of the engagement or during their appraisal cycle, In this fashion, the leader is highly respected and accepted by his or her teams.

Good supervisors deal with high performers in a different fashion, since these types of employees, while valuable and desirable, are often difficult to manage. The care, attention, praise, and other tools necessary to sustain performance require effort on the supervisor's part.

Adopting a cooperative team player attitude in your peer and superior relationships is a positive and logical pursuit to complement high performance and is the right thing to do. These leaders understand the behavior of the people they supervise and they learn who their people are and their needs and desires to be treated fairly and with dignity.

A Positive Attitude Towards Bonding

One of the most important skills successful supervisors master is the art of listening well. They identify the barriers that interfere with listening and practice effective ways of overcoming these barriers and becoming an excellent listener. These supervisors also master non-verbal communication and the differences in communication across cultures.

Many leaders in the organizations that talk passionately about being market-driven and customer-focused, however, have many times overlooked a crucial ingredient—the ability to listen well—and have caused their units to lose business. They did not always listen to what the customers had to say before telling them what they wanted.

Listening enables individuals to draw others into an interactive conversation in which they can ask perceptive questions, probe for reactions, and respond to those reactions appropriately, and this is the attitude of these leaders. However, most people are not natural listeners, let alone trained in the art of listening. That is probably because real listening involves letting go of your ego—temporarily subjugating one's own agenda in the interest of understanding another's message—and that is what is demanded from a good leader. The fact is, however, that most of us either do not hear the message at all, or hear it but misinterpret its meaning.

Listening involves several steps, some of which are demonstrated below.

- Hearing completely what is said without interrupting in the middle

- Interpreting what it really means without jumping to assumptions

- Responding and reacting in a positive way that shows that the message has been understood and is considered significant.

There is virtually no better way to create a favorable impression than by showing others that you are interested in their opinions and value them. Moreover, it is sometimes the only way that you can elicit attitudes and discover underlying needs—information that is crucial to satisfying others. This communication skill issued by these managers while communicating to all, whether it is with the customer, the stakeholders, or the project teams.

Effective Problem-Solving and Negotiation Skills

The savvy supervisor knows to look for and address problems before they escalate into major crises. They endeavor for win-win relationships with those they supervise, their customers, their peers, and those that supervise them.

Great leaders are those who are good problem solvers without panicking when faced with a problem. They have a systematic approach to solving problems. They understand that quick solutions or hastily-conceived plans seldom work and yield results that are short-term, at best. An effective leader is one who spends considerable time analyzing the problem from all possible angles, comes up with a solution, and at the same time, comes up with a backup plan, in case the original idea fails.

Some principles to bear in mind while entering into the problem-solving mode:

- Use effective listening.

- Maintain candid, open, and honest communication with the people involved.

- Do not resort to personal attacks.

- Observe how others are reacting to your conversations and change your style, if necessary, in a tricky situation.

Consider, for example, a situation that managers often face in people management. A new job applicant who is considered very good for the current opportunity is having serious discussions in terms of a higher salary. One option is to roll off the resource, which is never a wise decision and avoided by these managers. The best option that these managers choose is to offer them alternatives such as other reimbursements or location in another office, at the same time providing them with greater roles and responsibilities in the organization. The candidate might actually re-think his or her position and compromise on the salary for potential career growth. Instances like this show that negotiation is a useful tool for problem solving. Negotiations call for making decisions, but decision-making styles vary from one person to another and from one organization to another; regardless, the best negotiators are the best leaders.

Let me share with you an example where a high performing onsite team member, who had completed his assignment and who was supposed to join the offshore team, expressed his unwillingness to work at the offshore location. Negotiation skills played a very important role in this case, where the manager convinced him to return back onsite once the defined offshore for this assignment was over. The resource was given no choice except to accept this proposal else there were some negative consequences that were communicated positively. In this case, the factors of fear and the good negotiations both played very important roles in solving the problem.

However, there are different phases to negotiation. I am not discussing them in detail here, but here is a summary of some key ones:

- **Before the Negotiation**

  Perform analysis on the agenda of the discussion and get prepared. Beforehand, plan for alternatives to be put forth during the discussions and be prepared for surprises.

- **During the Negotiation**
  - ▸ **Setting the Right Tone**

    Have the right tone for the discussion to impress the participants and to draw the attention of the crowd. This also affects the language and the pronunciations we use.

  - ▸ **Exploring Underlying Needs**

    Discuss needs and get into the details to understand and make the participants understand the proposal or issue. A good listener and leader is always the one who lets other speak, interprets, and then decides.

  - ▸ **Developing Creative Alternatives**

    Work on alternatives towards the solution.

  - ▸ **Selecting the Best and Crafting an Agreement**

    Select the best alternative, get acceptance from the group, and then draft the agreement.

  - ▸ **Reviewing and Recapping the Agreement**

    Finally, recap the discussion and the agreement.

- **After the Negotiation**
  Implement the agreement and have periodic reviews as appropriate. Measure the output of the agreement and demonstrate the difference that has been made.

Leaders Show Leadership in Themselves

In the absence of a higher authority, leaders may be required to exercise judgment and/or authority that may exceed their charter, and they do so with common sense and address the situation logically. They, however, do so with the confidence within themselves that their supervisors will not negate their decisions and decisions and think out their decisions before exercising them. Successful supervisors anticipate most of the consequences of their decisions.

Decisions are at the heart of leader success and, at times, there are critical moments when they can be difficult, perplexing, and nerve-racking. However, well thought out and bold decisions are the safest and the best. Successful leaders are those who understand and implement this at the heart of their execution. Some decisions sometimes prove to be fatal, but a successful leader is one who does not fear making decisions and does not have "decidophobia." Fear of making the wrong decisions is well known to any responsible manager. Wherever you see a successful business, someone once made a courageous decision.

To be successful, you need the ability to apply physical and mental energies to one decision problem incessantly, without growing weary. A successful leader does not worry about making serious decisions when those decisions are honest and accurate to the best of his or her knowledge. Waiting to make decisions is like sitting on a rocking chair—it gives you something to do but does not get you anywhere.

231

Therefore, worrying about making a decision is a waste of time. A decision is something you can change in the future, if you must.

Successful leaders do not try to fool themselves and make a decision and then implement only one are two different parts of it. A successful leader is one who has the capability to implement the decisions he or she has made. The measure of success is not whether they have a tough decision to make but, whether it is the same decision they had made once before, but have failed to implement in the past.

Successful supervisors have a clear plan of what has to be achieved, and determine what will be the result of the decision made today. They operate with complete confidence that their anticipation of the consequences is perfect.

These supervisors combine their analytical and leadership qualities to anticipate future directions because they have already planned their work, anticipated problems, and have planned for workarounds. Their subordinates also demonstrate higher confidence in these supervisors as they know that these supervisors know what is going to happen.

References

[1] Agile Alliance/Manifesto. <http://www.agilealliance.org>.

[2] Agile Enterprise. <http://www.e-architects.com/AE/>.

[3] Agile Modeling. <http://agilemodeling.com/>.

[4] "Agile Software Development: Business of Innovation." <http://www.adaptivesd.com/Articles/ IEEEArticle1Final.pdf>.

[5] "Agility Counts." *Economist*. (2001): 11.

[6] Aoyama, Mikio. "Web-Based Agile Software Development," *IEEE Software*. Nov/Dec 1998.

[7] Auer, Ken and Roy Miller. *Extreme Programming Applied: Playing to Win*. Addison-Wesley, 2002.

[8] Baird, Stewart, Paul Peterson, and D. Voss. "XP Practices in Action." *In Sam's Teach Yourself Extreme Programming in 24 Hours*. Indianapolis: Sam's Publishing, 2002.

[9] Beedle, Mike and Ken Schwaber. *Agile Software Development with SCRUM*. Pearson Technology Group, 2002.

[10] Boehm, B. "Get Ready for Agile Methods with Care," *IEEE Computer.* January 2002.

[11] Boehm, B., A. Egyed, J. Kwan, D. Port, A. Shah, and R. Madachy, "Using the Win-Win Spiral Model: A Case Study," *IEEE Computer.* July 1998. <http://sunset.usc. edu/publications/TECHRPTS/1998/usccse98-512/ usccse98-512.pdf>.

[12] Cockburn, Alistair. *Agile Software Development.* Addison Wesley, 2002.

[13] Cockburn, A. and J. Highsmith. "Agile Software Development: The People Factor," *IEEE Computer.* Nov. 2001.

[14] Crystal/Adaptive Software Development. <http://agile.csc. ncsu.edu/crystal.html>.

[15] Fowler, M. and J. Highsmith. "Agile Manifesto," *Software Development.* August 2001.

[16] Highsmith, Jim. *Agile Software Development Ecosystems.* Addison-Wesley, 2003.

[17] Jacobson, I. "A Resounding 'Yes' to Agile Processes—but also to More," *Cutter IT Journal.* Vol. 15, No. 1, (2001: 18-24).

[18] Jeffries, R. "Card Magic for Managers," *Software Development*, December 2000.

[19] Martin, Robert C. and James W. Newkirk. *Extreme Programming in Practice.* Addison- Wesley, 2001.

[20] Muller, Mathias. "Are Reviews an Alternative to Pair Programming?" *Empirical Software Engineering*, December 2004.

[21] Principles of the Agile Alliance. <http://www.agilealliance. org>.

[22] VanDeGrift, Tammy. "Coupling pair programming and writing: learning about students' perceptions and processes," Proceedings of the 35th SIGCSE Technical Symposium on *Computer Science Education*, March 2004.

[24] Wake, William C. *Extreme Programming Explored*. Addison-Wesley, 2001.

About the Author

Upadrista Venkatesh began compiling technical encyclopedias as a high school student nearly twenty years ago. Born in India in 1976, Venkat studied mathematics and computers at Andhra University and received his bachelor's degree in 1996. In 1995, Venkat took the vast collection of experimental facts from various sources and the research that he had been accumulating since his teenage years and began to  deploy them on the early Internet. These pioneering efforts at organizing and presenting his ideas on technical designs defined a paradigm that other large-scale informational projects on the web have subsequently followed. In the late 90s, he shifted his career and perspective from the technical venue to management. Since then, he has been experimenting with and researching different management principles in a successful managerial career.

Upadrista Venkatesh is currently managing the delivery of multiple accounts for one of India's top IT service providers. He has been an advisor to a number of e-business initiatives and has had a successful career in setting up PMOs and competency centers, aligning new verticals, etc. in his the IT industry. He has extensive experience working with customers and has been successful in all of his major assignments, earning the customers' confidence. He also has worked in vendor management, being motivated to see himself succeed in the goal along with the vendor organization. His experience moves across product and project management. He has worked in a blended mixture of the onsite/offshore model and through his experience, has helped some customers move their strategies from the onsite model to the onsite/offshore model. He has a proven successful record of accomplishment as the advisor for major customers transforming their project models from the traditional life cycle model to the Agile development model.

He has dealt with large project agreements bringing accounts from the conceptual stage to million–dollar programs and has demonstrated great expertise in pre-sales as well.

His current research and interests focus on the management of strategic changes in an organization seeking to move to the Agile development approach.

You can reach him on his number 91-9849643233 or on his e-mail @ Upadrista@yahoo.com.

Managing Agile Projects

Are you being asked to manage a project with unclear requirements, high levels of change, or a team using Extreme Programming or other Agile Methods?

If you are a project manager or team leader who is interested in learning the secrets of successfully controlling and delivering agile projects, then this is the book for you.

From learning how agile projects are different from traditional projects, to detailed guidance on a number of agile management techniques and how to introduce them onto your own projects, this book has the insider secrets from some of the industry experts – the visionaries who developed the agile methodologies in the first place.

ISBN: 1-895186-11-0   (paperback)
ISBN: 1-895186-12-9   (PDF ebook)

http://www.mmpubs.com

Surprise! Now You're a Software Project Manager

It's late Friday afternoon and you have just been told by your boss that you will be the project manager for a new software development project starting first thing on Monday morning. Congratulations! Now, if only you had taken some project management training...

This book was written as a crash course for people with no project management background but who still are expected to manage a small software development project. It cuts through the jargon and gives you the basics: practical advice on where to start, what you should focus on, and where you can cut some corners. This book could help save your project... and your job!

ISBN: 1-895186-75-7 (paperback)
ISBN: 1-895186-76-5 (PDF ebook)

http://www.mmpubs.com/surprise

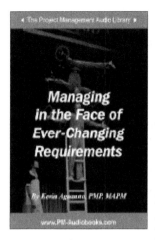

Managing in the Face of Ever-Changing Requirements

1000% over budget. 2 years late. We've all heard about projects where the original budget and schedule were exceeded by orders of magnitude. How could this happen given competent project management? Whether in government or private industry, project managers face a huge challenge when confronted with continually-changing requirements. How can you plan when the WBS keeps changing? Traditional project management approaches suggest that we take a snapshot of the requirements as a baseline, and then use change control in an attempt to minimize the impact of any shifting requirements. Sometimes, however, these changes are a reality that the project sponsor has to accommodate. The new Agile Project Management methods help deal with these situations.

Listeners of this session will learn the founding principles and techniques of Agile Project Management with examples from real-world projects that used these methods to control changing requirements.

ISBN: 1-895186-32-3 (Audio CD)

http://www.PM-Audiobooks.com

Agile Project Management Using Scrum

In recent years, the Internet revolution has caused a shift in how fast technology is developed and marketed. We have seen the appearance of "Web Years" as a measure of time, and the widespread adoption of Rapid Application Development (RAD) as a standard software development method used in even our largest organizations.

There has been a parallel shift in how projects are managed. First appearing in software development pro-jects, Agile Development methods are now a very hot topic in software development conferences and magazines. These are methods that stress the speed of development and close interaction with the customer over traditional, more bureaucratic, practices.

This recording will outline the underlying principles of Agile Development and details of how it differs from traditional development projects. Then, using an agile project management method called Scrum, it will illustrate how agile management methods used in software development may be extended to projects from other application areas outside of I/T. Listeners will come away from the session with a high-level understanding of the Agile Development philosophy and how it differs from traditional development approaches, enough of an understanding of Scrum to be able to determine if and how it could be implemented on a project, and a list of resources for further information on Agile Development and Scrum.

ISBN: 1-895186-13-7   (Audio CD)

http://www.PM-Audiobooks.com

Agile Database Development

Data is clearly an important aspect of software-based systems; yet, many agile development teams are struggling to involve data professionals within their projects. The Agile Data (AD) methodology defines a philosophical framework for data-oriented activities within agile projects, defining ways that application developers and data professionals can work together effectively; however, philosophy isn't enough — you also need proven techniques which support those philosophies.

This recording presents techniques for agile database development, including database refactoring, Agile-Model Driven Development (AMDD), Test-Driven Design (TDD), and environment/tool strategies.

ISBN: 1-895186-33-5   (Audio CD)

http://www.PM-Audiobooks.com

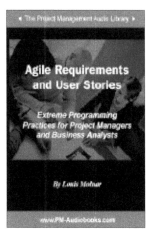

Agile Requirements and User Stories: Extreme Programming Practices for Project Managers and Business Analysts

Many organizations are starting to reap rewards from adding Agile methods to their development practices. Learn what Project Managers and Business Analysts need to know about Extreme Programming (XP), User Stories and "The Planning Game." Specifically, this recording will give practical Agile tips and tricks including how to specify User Stories to effectively drive out your Client's true business requirements.

In this recording, you will:

- Understand how User Stories can be an effective technique for gathering user requirements.

- Learn tips & techniques on how to do User Stories effectively.

- Know how detailed User Stories need to be specified.

- Know exactly how many User Stories are required.

- Learn the PM's and BA's role in an XP development environment.

ISBN: 1-895186-45-5   (Audio CD)

http://www.PM-Audiobooks.com

Managing Smaller Projects: A Practical Approach

So called "small projects" can have potentially alarming consequences if they go wrong, but their control is often left to chance. The solution is to adapt tried and tested project management techniques.

This book provides a low overhead, highly practical way of looking after small projects. It covers all the essential skills: from project start-up, to managing risk, quality and change, through to controlling the project with a simple control system. It cuts through the jargon of project management and provides a framework that is as useful to those lacking formal training, as it is to those who are skilled project managers and want to control smaller projects without the burden of bureaucracy.

Read this best-selling book from the U.K., now making its North American debut. IEE Engineering Management praises the book, noting that "Simply put, this book is about helping people control smaller projects in a logical and effective way, and making the process run smoothly, and is indeed a success in achieving that goal."

Available in print format. Order from your local bookseller, Amazon.com, or directly from the publisher at
http://www.mmpubs.com

Winston Churchill: The Agile Project Manager

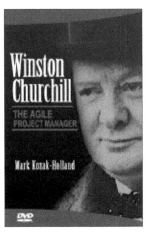

Today's pace of change has reached unprecedented levels only seen in times of war. As a result, project management has changed accordingly with the pressure to deliver and make things count quickly. This recording looks back at a period of incredible change and mines lessons for Project Managers today.

In May 1940, the United Kingdom (UK) was facing a dire situation, an imminent invasion. As the evacuation of Dunkirk unfolded, the scale of the disaster became apparent. The army abandoned 90% of its equipment, the RAF fighter losses were deplorable, and over 200 ships were lost.

Winston Churchill, one of the greatest leaders of the 20th century, was swept into power. With depleted forces and no organized defense, the situation required a near miracle. Churchill had to mobilize quickly and act with agility to assemble a defense. He had to make the right investment choices, deploy resources, and deliver a complete project in a fraction of the time. This recording looks at Churchill as an agile Project Manger, turning a disastrous situation into an unexpected victory.

ISBN: 1-895186-50-1   (Audio CD)
ISBN: 1-897326-38-6   (DVD)

http://www.PM-Audiobooks.com

Churchill's Adaptive Enterprise: Lessons for Business Today

This book analyzes a period of time from World War II when Winston Churchill, one of history's most famous leaders, faced near defeat for the British in the face of sustained German attacks. The book describes the strategies he used to overcome incredible odds and turn the tide on the impending invasion. The historical analysis is done through a modern business and information technology lens, describing Churchill's actions and strategy using modern business tools and techniques. Aimed at business executives, IT managers, and project managers, the book extracts learnings from Churchill's experiences that can be applied to business problems today. Particular themes in the book are knowledge management, information portals, adaptive enterprises, and organizational agility.

Eric Hoffer Book Award (2007) Winner

ISBN: 1-895186-19-6   (paperback)
ISBN: 1-895186-20-X   (PDF ebook)

http://www.mmpubs.com/churchill

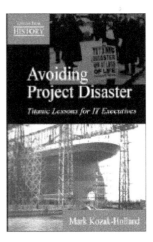

Avoiding Project Disaster: Titanic Lessons for IT Executives

Imagine you are in one of Titanic's lifeboats. As you look back at the wreckage site, you wonder what could have happened. What were the causes? How could things have gone so badly wrong?

Titanic's maiden voyage was a disaster waiting to happen as a result of the compromises made in the project that constructed the ship. This book explores how modern executives can take lessons from a nuts-and-bolts construction project like Titanic and use those lessons to ensure the right approach to developing online business solutions. Looking at this historical project as a model will prove to be incisive as it cuts away the layers of IT jargon and complexity.

Avoiding Project Disaster is about delivering IT projects in a world where being on time and on budget is not enough. You also need to be up and running around the clock for your customers and partners. This book will help you successfully maneuver through the ice floes of IT management in an industry with a notoriously high project failure rate.

ISBN: 1-895186-73-0   (paperback)

Also available in ebook formats.

http://www.mmpubs.com

Titanic Lessons for IT Projects

Titanic Lessons for IT Projects analyzes the project that designed, built, and launched the ship, showing how compromises made during early project stages led to serious flaws in this supposedly "perfect ship." In addition, the book explains how major mistakes during the early days of the ship's operations led to the disaster. All of these disasterous compromises and mistakes were fully avoidable.

Entertaining and full of intriguing historical details, this companion book to Avoiding Project Disaster: Titanic Lessons for IT Executives helps project managers and IT executives see the impact of decisions similar to the ones that they make every day. An easy read full of illustrations and photos to help explain the story and to help drive home some simple lessons.

ISBN: 1-895186-26-9   (paperback)

Also available in ebook formats.

http://www.mmpubs.com/titanic

The Elusive Lean Enterprise: Why Companies Struggle to Get the Full Benefits from Lean

In today's fast-paced and volatile business environment, customers are demanding increased flexibility and lower cost, and companies must operate in a waste-free environment to maintain a competitive edge and grow margins. Lean Enterprise is the process that companies are adopting to provide superior customer service and improve bottom line performance.

Are you contemplating Lean Enterprise for your manufacturing or office facility? Are you already implementing Lean, but dissatisfied with the speed of change? Do your employees think that Lean is just the new flavor of the month? Are you being forced to go Lean by your customers?

This book is designed to help guide you through the Lean transformation and avoid the pitfalls. Find out why many companies are failing to live up to the promise of Lean, and why there may be alternatives to outsourcing or going offshore.

ISBN: 1-897326-64-5   (paperback)
ISBN: 1-897326-65-3   (hardcover)
ISBN: 1-897326-66-1   (Adobe PDF eBook)

http://www.mmpubs.com/

Are People Finding Your Meetings Unproductive and Boring?

Turn ordinary discussions into focused, energetic sessions that produce positive results.

If you are a meeting leader or a participant who is looking for ways to get more out of every meeting you lead or attend, then this book is for you. It's filled with practical tips and techniques to help you improve your meetings.

You'll learn to spot the common problems and complaints that spell meeting disaster, how people who are game players can effect your meeting, fool-proof methods to motivate and inspire, and templates that show you how to achieve results. Learn to cope with annoying meeting situations, including problematic participants, and run focused, productive meetings.

ISBN: 1-897326-15-7   (paperback)

Also available in ebook formats.

http://www.mmpubs.com/

Want to Get Ahead in Your Career?

Do you find yourself challenged by office politics, bad things happen-ing to good careers, dealing with the "big cheeses" at work, the need for effective networking skills, and keeping good working relation-ships with coworkers and bosses? *Winning the Rat Race at Work* is a unique book that provides you with case studies, interactive exercises, self-assessments, strategies, evaluations, and models for overcoming these workplace challenges. The book illustrates the stages of a career and the career choices that determine your future, empowering you to make positive changes.

Written by Peter R. Garber, the author of *100 Ways to Get on the Wrong Side of Your Boss*, this book is a must read for anyone interested in getting ahead in his or her career. You will want to keep a copy in your top desk drawer for ready reference whenever you find yourself in a challenging predicament at work.

ISBN: 1-895186-68-4   (paperback)

Also available in ebook formats. Order from your local bookseller, Amazon.com, or directly from the publisher at **http://www.mmpubs.com/rats**

Need More Help with the Politics at Work?

100 Ways To Get On The Wrong Side Of Your Boss (And Strategies to Prevent You from Getting There!) was written for anyone who has ever been frustrated by his or her working relationship with the boss—and who hasn't ever felt this way! Bosses play a critically important role in your career success and getting on the wrong side of this important individual in your working life is not a good thing.

Each of these 100 Ways is designed to illustrate a particular problem that you may encounter when dealing with your boss and then an effective strategy to prevent this problem from reoccurring. You will learn how to deal more effectively with your boss in this fun and practical book filled with invaluable advice that can be utilized every day at work.

Written by Peter R. Garber, the author of *Winning the Rat Race at Work*, this book is a must read for anyone inter-ested in getting ahead. You will want to keep a copy in your top desk drawer for ready reference whenever you find yourself in a challenging predicament at work.

ISBN: 1-895186-98-6   (paperback)

Also available in ebook formats. Order from your local bookseller, Amazon.com, or directly from the publisher at **http://www.InTroubleAtWork.com**

Lessons from the Ranch for Today's Business Manager

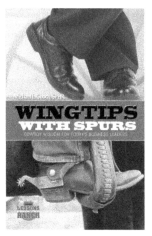

The lure of the open plain, boots, chaps and cowboy hats makes us think of a different and better way of life. The cowboy code of honor is an image that is alive and well in our hearts and minds, and its wisdom is timeless.

Using ranch based stories, author Michael Gooch, a ranch owner, tells us how to apply cowboy wisdom to our everyday management challenges. Serving up straight forward, practical advice, the book deals with issues of dealing with conflict, strategic thinking, ethics, having fun at work, hiring and firing, building strong teams, and knowing when to run from trouble.

A unique (and fun!) approach to management training, Wingtips with Spurs is a must read whether you are new to management or a grizzled veteran.

ISBN: 1-897326-88-2   (paperback)

Also available in ebook formats. Order from your local bookseller, Amazon.com, or directly from the publisher at **http://www.mmpubs.com**

 **The Project Management Audio Library**

In a recent CEO survey, the leaders of today's largest corporations identified project management as the top skillset for tomorrow's leaders. In fact, many organizations place their top performers in project management roles to groom them for senior management positions. Project managers represent some of the busiest people around. They are the ones responsible for planning, executing, and controlling most major new business activities.

Expanding upon the successful *Project Management Essentials Library* series of print and electronic books, Multi-Media Publications has launched a new imprint called the *Project Management Audio Library*. Under this new imprint, MMP is publishing audiobooks and recorded seminars focused on professionals who manage individual projects, portfolios of projects, and strategic programmes. The series covers topics including agile project management, risk management, project closeout, interpersonal skills, and other related project management knowledge areas.

This is not going to be just the "same old stuff" on the critical path method, earned value, and resource levelling; rather, the series will have the latest tips and techniques from those who are at the cutting edge of project management research and real-world application.

www.PM-Audiobooks.com

www.ingramcontent.com/pod-product-compliance
Lightning Source LLC
Chambersburg PA
CBHW071108050326
40690CB00008B/1158